THROUGH THE LOOKING GLASS

ISSUES OF PSYCHOLOGICAL WELL-BEING IN CAPTIVE NONHUMAN PRIMATES

THROUGH THE LOOKING GLASS

ISSUES OF PSYCHOLOGICAL WELL-BEING IN CAPTIVE NONHUMAN PRIMATES

EDITED
BY
MELINDA A. NOVAK
AND
ANDREW J. PETTO

American Psychological Association
Washington, DC

Published by
American Psychological Association
1200 Seventeenth Street, NW
Washington, DC 20036

Copies may be ordered from
APA Order Department
P.O. Box 2710
Hyattsville, MD 20784

Designed by Paul M. Levy (cover artwork by Sarah Petto)
Typeset by Harper Graphics, Waldorf, MD
Printed by BookCrafters
Technical editing and production coordinated by
Deborah Segal
Copyediting by Cynthia L. Clark

Library of Congress Cataloging-in-Publication Data

Through the looking glass : issues of psychological well-being in captive nonhuman primates / edited by Melinda A. Novak and Andrew J. Petto.
p. cm.
Includes bibliographical references and index.
ISBN 1-55798-087-X (acid-free paper)
1. Primates as laboratory animals—Congresses. 2. Primates—Psychology—Congresses. 3. Captive wild animals—Psychology—Congresses. 4. Animal welfare—Congresses. I. Novak, Melinda A. II. Petto, Andrew J.
SF407.P7T49 1991
636′.98—dc20 90-27583
CIP

Printed in the United States of America
First edition

CONTENTS

CONTRIBUTORS

Patricia Alford, *University of Texas, M. D. Anderson Cancer Center*
Susan A. Beckley, *Department of Zoology, University of Massachusetts at Amherst*
B. Taylor Bennett, *Biologic Resource Lab, University of Illinois at Chicago*
Molly Bloomsmith, *University of Texas, M. D. Anderson Cancer Center*
Steve Carroll, *Incurably Ill for Animal Research, Bridgeview, IL*
Christopher L. Coe, *University of Wisconsin*
W. Jean Dodds, *New York State Department of Health, Wadsworth Center*
Karla H. Drewsen, *University of Massachusetts at Amherst*
Joseph M. Erwin, *Diagnon Corporation, Washington, DC*
Ronald D. Hunt, *Harvard Medical School, New England Regional Primate Research Center*
M. Kay Izard, *Research Triangle Institute*
Lorna Johnson, *Harvard Medical School, New England Regional Primate Research Center*
Michale Keeling, *University of Texas, M. D. Anderson Cancer Center*
Frederick A. King, *Yerkes Regional Primate Research Center, Emory University*
Edward A. Leonard, *Animal Rescue League of Boston*
Scott Line, *California Primate Research Center, University of California at Davis*
Franklin M. Loew, *Tufts University School of Veterinary Medicine*
Hal Markowitz, *Department of Biology, San Francisco State University*
Kathleen N. Morgan, *California Primate Research Center, University of California at Davis*
Melinda A. Novak, *University of Massachusetts at Amherst*
Andrew Petto, *Harvard Medical School, New England Regional Primate Research Center*
Viktor Reinhardt, *Wisconsin Regional Primate Research Center, University of Wisconsin*
Rosalind M. Rolland, *Harvard Medical School, New England Regional Primate Research Center*

Leonard Rosenblum, *Department of Psychiatry, SUNY Health Science Center at Brooklyn*

Gene P. Sackett, *Regional Primate Research Center, University of Washington*

Dale Schwindaman, *United States Department of Agriculture, Animal and Plant Health Inspection Service*

Prabhat K. Sehgal, *Harvard Medical School, New England Regional Primate Research Center*

Charles T. Snowdon, *University of Wisconsin*

Sharon Strong, *Department of Biology, San Francisco State University*

Stephen J. Suomi, *National Institutes of Child Health and Development*

Frans B. M. de Waal, *Wisconsin Regional Primate Research Center, University of Wisconsin*

Thomas L. Wolfle, *National Research Council, Washington, DC*

FOREWORD

Federal research agencies stopped most support of investigator-initiated "state-of-the-art" research conferences in scientific psychology over a decade ago. Yet during this period, scientific psychology has continued to grow, as well as to diversify into many new areas. Thus, there have been relatively few opportunities for investigators in new and promising research areas to convene in special settings to discuss their findings.

As part of its continuing effort to enhance the dissemination of scientific knowledge in psychology, the American Psychological Association (APA) undertook a number of initiatives designed to foster scientific research and communication. In particular, the APA Science Directorate established the Scientific Conferences Program in 1988.

The APA Scientific Conferences Program provides university-based psychological researchers with seed monies essential to organizing specialty conferences on critical issues in basic research, applied research, and on methodological issues in psychology. Deciding which conferences to support involves a competitive process. An annual call for proposals is issued by the APA Science Directorate to solicit conference ideas. Proposals from all areas of psychological research are welcome. These are then reviewed by qualified psychologists, who forward substantive suggestions and funding recommendations to the Science Directorate. At each stage, the criteria used to determine which conferences to support include relevance, timeliness, comprehensiveness of the topics, and qualifications of the presenters. In 1988, seven conferences were funded through this program of the Science Directorate, and seven conferences were funded in 1989. We expect to support several more each year, at an annual program expense of $90,000 to $100,000.

This program has two major goals. The first is to provide, by means of the conferences, a broad view of specific topics (and when appropriate, to provide for interdisciplinary participation). The second goal is to assure timely dissemination of the findings presented, by publishing carefully crafted scholarly volumes based, in part, on the conferences. The information thus reaches the broader psychological and scientific communities, as well as the audiences at the conferences. Psychology and related fields are thereby benefited from the most current research on a given topic.

This volume presents findings reported at the September 1988 conference, "The Psychological Well-being of Captive Primates." The impetus for this conference was a 1985 amendment to the Animal Welfare Act, which required researchers to "provide a physical environment adequate to promote the psychological well-being of primates" while in captivity. As federal policy-makers attempted to draft regulations imposing requirements on the research community, it became clear that many research questions about "psychological well-being" had yet to be answered. In this volume, researchers and policy-makers present diverse opinions about the current state of knowledge on "psychological well-being," the meaning of the legislation, and the implications and effects it will have on the research community.

This volume represents what we at the American Psychological Association believe are a number of exceptional volumes providing readers with a broad sampling of the diverse and outstanding psychological research now being conducted by our colleagues. We hope you will enjoy and be stimulated by this book and others to come.

A list of the conferences funded through this program follows:

Researching Community Psychology: Integrating Theories and Methodologies, September 1988
Psychological Well-being in Nonhuman Captive Primates, September 1988
Psychological Research on Organ Donation, October 1988
Arizona Conference on Sleep and Cognition, January 1989
Socially Shared Cognition, February 1989
The Role of Experience in Modifying Taste and Its Effects on Feeding, April 1989
Perception of Structure, May 1989
Suggestibility of Children's Recollections, June 1989
Best Methods for the Analysis of Change, October 1989
Conceptualization and Measurement of Organism–Environment Interaction, November 1989
Cognitive Bases of Musical Communication, April 1990
Conference on Hostility, Coping/Support, and Health, November 1990
Psychological Testing of Hispanics, February 1991
Study of Cognition: Conceptual and Methodological Issues, February 1991

Lewis P. Lipsitt, PhD
Executive Director
Science Directorate, APA

Virginia E. Holt
Manager, Scientific Conferences Program
Science Directorate, APA

PREFACE

This book is an outgrowth of a conference, *The Psychological Well-being of Captive Primates*, which was held in September of 1988 at Harvard Medical School in Boston, Massachusetts. The catalyst for the conference was the 1985 amendment to the Animal Welfare Act, which specifically required primatologists to "provide a physical environment adequate to promote the psychological well-being of primates." The purposes of the conference were to identify and define the term "psychological well-being" for nonhuman primates, to determine ways in which psychological well-being might be enhanced, to consider the economic impact of the 1985 amendment, and to evaluate the effect of such legislation on the research enterprise. Diverse viewpoints were provided by researchers, U.S. Department of Agriculture officials, veterinarians, and animal welfare groups. Given the crucial nature of the legislation, a number of institutions came forward to sponsor the conference, namely, the University of Massachusetts—Amherst, Tufts University, and Harvard Medical School. We are indebted to these institutions for their financial contributions and staff support. The conference was also supported by a substantial grant from the American Psychological Association, without which this public forum could not have taken place.

We would like to extend our appreciation to all the people who made this book possible, but whose names are not likely to appear anywhere else. We are indebted to our students and colleagues, Sue Beckley, Karla Drewsen, Michele Lareau, Kathy Parks, Arlene Walsh, and Lyna Watson, for their thoughtful and insightful discussions of the issues presented in this book. For staff services of various kinds, we thank Pat Kane, Helen Pluta, and Sydney Fingold. Very special thanks go to our typist, Debbie Brosseau, and indexer, Julie Goldberg. Finally, we would like to acknowledge our spouses, Jerrold Meyer and Sarah Petto, whose support and faith helped us to achieve our goal.

Melinda A. Novak
Andrew J. Petto

CHAPTER 1

PERSPECTIVES ON PSYCHOLOGICAL WELL-BEING IN CAPTIVE PRIMATES: THROUGH THE LOOKING GLASS

MELINDA A. NOVAK AND ANDREW J. PETTO

> Then she began looking about, and noticed that what could be seen from the old room was quite common and uninteresting, but that all the rest was as different as possible.
>
> —Lewis Carroll,
> *Through the Looking Glass*

It is easy for us, as humans, to see ourselves reflected in the lives of our nonhuman primate cousins. Given the close biological and evolutionary relationship that we share, it would be surprising only if this were not so. But beyond our own reflection in the glass is a group of over 200 species that is behaviorally and biologically diverse. Understanding the complex pattern of similarities and differences among human and nonhuman primates presents a formidable challenge. Only when we have explored what Lewis Carroll called "all the rest" can we understand the meaning of the reflection in the looking glass and why it sometimes yields unusual images.

Within the last decade, a considerable debate has been brewing over the use of animals such as primates in biomedical and behavioral research. Although some of the arguments involve issues of morality and ethics, a significant portion of the debate concerns how primates are housed and maintained in research laboratories. Animal-rights activists frequently paint a distorted picture of scientists in the conduct of their research. This picture features uncaring, indifferent researchers subjecting helpless animals held hostage in bleak environments to unnecessary and irrelevant research. Despite the overwhelming inaccuracy of this view and the indignation it arouses, scientists have been largely ineffective in refuting such claims in the popular press. Thus, a certain ambivalence appears to have developed in recent years, even among the majority of the public that favors the use of animals in research. It is not surprising, given this climate of opinion, that scientists face increased federal regulation of scientific endeavors.

In 1985, Congress amended the Animal Welfare Act to include a provision that was to have a significant impact on the primate research community. This new provision specifically required primate researchers to provide "a physical environment adequate to promote psychological well-being in primates" [Animal Welfare Act, 1985 (Section 13)]. This phraseology created quite a stir in the scientific community. It went beyond the well-established boundaries of humane care and treatment into an area where considerably less was known and where the risk of "creeping anthropomorphism" appeared to be very great. On the other hand, to some members of the public (especially those involved in animal rights organizations), the new legislation did not go far enough.

In the years since the new legislation was passed, regulators and researchers alike have grappled with the problems of interpreting psychological well-being and of devising a set of workable regulations that meet the intent of the amendment. At least two issues of disagreement has arisen between researchers and regulators. The first problem is one of orientation. The question posed here is whether the regulations should be based on structure (e.g., cage size) or function (e.g., primate responsivity), or as some characterize it, engineering as opposed to performance standards. A second but interrelated problem is one of generality as it pertains to diversity (i.e., to what extent will the regulations have broad applicability across the many different arboreal and terrestrial primate species residing in laboratories?)

These are two of the recurrent themes that emerged from the conference and are explored by the contributors in the chapters that follow. We have divided these contributions into six sections to provide a guide to the main issues presented in each chapter and to the context in which they were addressed.

PART ONE. HISTORICAL PERSPECTIVES ON PSYCHOLOGICAL WELL-BEING LEGISLATION

The first section is concerned with an historical view of animal research and animal welfare. In chapter 2, Loew explores some of the public concerns, both current and historic, that have led to major public policy changes regarding the use of animals in research. He stresses the need for scientists to listen to alternative viewpoints. Hunt's historical account (chapter 3) emphasizes the primary role that science has played and must continue to play in the development of appropriate standards for the humane care and use of laboratory animals. As he notes, scientific evidence forms the necessary core upon which effective guidelines and regulations are typically based. In chapter 4, Schwindaman presents the views and aims of the United States Department of Agriculture (USDA) with respect to the recent amendments.

PART TWO. DEFINING PSYCHOLOGICAL WELL-BEING

Terminology is one of the primary concerns of the Animal Welfare Act (1985). A term like "psychological well-being," which forms the core of the proposed regulations (USDA, August 15, 1990), is extremely difficult to define and measure adequately in human or nonhuman primates and therefore poses a significant problem to researchers and regulators alike. In chapter 5, Sackett examines the human literature for models of psychological well-being and evaluates their potential application to nonhuman primates. He then raises the issue of the relationship between structure and function. Using deinstitutionalization as an example, Sackett demonstrates that successful programs are those that put function first and then develop a structure to serve the function. He draws a compelling parallel between the problems arising from deinstitutionalization and the problems that scientists face with the new regulations.

Rosenblum (chapter 6) examines the role of objective and subjective factors in assessing psychological well-being in nonhuman primates. He focuses first on empathy, or anthropomorphism, and warns of the dangers when empathy becomes projection. Rosenblum then discusses the importance of more objective criteria such as naturalness, allometric relationships, and environmental control in fostering psychological well-being in captive primates.

In chapter 7, Suomi and Novak describe their strategy for measuring psychological well-being in laboratory primates. Once such measurement has been established, researchers can proceed to evaluate the effectiveness of various environmental structures in promoting psychological well-being. Suomi and Novak note that general principles for promoting psychological well-being are not likely to emerge. Rather, different structures or features will be required to promote the psychological well-being of different species, of different sex and age-class groups within a species, and of individuals of different temperament.

Keeling, Alford, and Bloomsmith take a different approach in chapter 8, the final chapter of this section. They argue that although theoretical discussions of psychological well-being are valuable, a "bias for action" is also necessary. Too much time can be spent in planning without accomplishing important goals. Even when available information is incomplete or inconclusive, critical decision-analysis procedures (similar to clinical trials) can be used to design programs for promoting psychological well-being in captive primates. Such programs are to be evaluated, updated, and revised on a regular schedule.

PART THREE. POTENTIAL PARAMETERS OF PSYCHOLOGICAL WELL-BEING

In the third section, various parameters of psychological well-being are examined. In chapter 9, de Waal paints a compelling portrait of the social needs of primates. He argues that primates such as apes and Old World monkeys have extremely sophisticated social systems as evidenced by their political strategies, reconciliation behavior, and kinship patterns. According to de Waal, social companionship is more than just the proximity of other warm bodies. Despite some potential costs (e.g., aggression, wounding), providing a social milieu may be the single best method for promoting psychological well-being in nonhuman primates.

Although Coe (chapter 10) clearly agrees that companionship is important for primates, he cautions about the effects of such social interactions. For example, animals that previously coexisted may become markedly aggressive to one another when a member of the opposite sex is introduced into the same room (not cage). Since movement of animals to different cages in different rooms is a common practice at many institutions, Coe argues that colony management problems will increase with the expansion of social housing. He also provides some startling findings indicating that old monkeys paired with infants and juveniles may actually show immune suppression, in contrast to old monkeys that remain by themselves.

In chapter 11, Johnson, Petto, and Sehgal use reproduction and parental care as potential parameters of psychological well-being in captive tamarins. In their studies, they have varied the complexity of both the physical and social environment and examined the effects of these manipulations on reproduction and parental behavior. Although the degree of social complexity is positively correlated with reproduction, survivorship, and quality of parental care, changes in the physical environment do not affect these variables in the expected manner.

In the final chapter of this section, Snowdon demonstrates that control over the environment is an extremely important aspect of the psychological well-being of captive primates. He further argues that such control can be achieved when primates are housed in naturalistic environments designed to simulate features of natural habitats. He describes techniques that enable him to measure learning performance in individuals and to assess hormonal status in females during pregnancy and lactation while the animals remain in their naturalistic setting.

PART FOUR. IMPLICATIONS OF THE NEW AMENDMENTS FROM AN ADMINISTRATIVE AND A VETERINARY PERSPECTIVE

The authors of this section focus on the development of institutional programs to promote psychological well-being in captive nonhuman primates. Wolfle (chapter 13) describes the importance of the institutional structure and the development of administrative responsibility. He notes that the person in authority must have control over work personnel, money, space, and other resources if an effective program is to emerge. The task cannot be delegated merely to the veterinarian. Ultimately, effective programs will be those that are allowed to change with time. As Wolfle suggests, flexibility must be present at both a regulatory and an administrative level.

Rolland (chapter 14) discusses the considerable challenge that clinical veterinarians will face in implementing the new regulations. In addition to noting the lack of an adequate definition of *psychological well-being*, Rolland indicates that many veterinarians are not sufficiently trained in the area of primate behavior. She then describes the impact of the new social-housing regulations from the perspective of the veterinarian.

PART FIVE. PROCEDURES FOR PROMOTING PSYCHOLOGICAL WELL-BEING IN NONHUMAN PRIMATES

Although research on the topic of psychological well-being is neither abundant nor always clearcut, there is growing evidence that some techniques are better than others. In this section, several researchers talk about the methodology they use to promote psychological well-being in captive primates. Izard, for example, discusses, in chapter 15, the indoor and outdoor environments that have been created for different species of prosimian primates at the Duke Primate Center. She focuses on features of the physical environment (e.g., vine tangles, nest boxes), diet, and social groupings.

Reinhardt (chapter 16) describes an environmental enrichment program for individually housed rhesus monkeys instituted at the Wisconsin Regional Primate Research Center. Here the emphasis is on the addition of manipulable objects such as branches, as well as on socializing animals that had been previously kept alone for long periods of time. Reinhardt provides compelling evidence that individually housed monkeys can be successfully paired together and that the presence of a social companion does not interfere with many biomedical procedures.

In chapter 17, Line, Markowitz, Morgan, and Strong focus on individually housed macaques. Using behavioral, hormonal, and cardiac measures, they show that monkeys continue to react to certain events, such as cage cleaning, even though they presumably are habituated to them. When considering various types of environmental enrichment, the authors report that increasing the cage size has relatively little effect on the monkeys in comparison to the provision of a music-feeder box over which the monkeys have control.

In the final chapter in this section, Erwin focuses on cage size and enrichment devices (e.g., toys). He presents strong evidence demonstrating that in determining optimal cage size "more is not always better" and that intuition may not always provide a rational basis for determining what is beneficial for captive primates. He notes that there is considerable interspecific variation in response to enrichment devices. Macaques and chimpanzees manipulate and play with objects, while patas and vervet monkeys show little interest. Thus, general prescriptions for promoting psychological well-being will have to be modified for different species.

PART SIX. PUBLIC REACTION TO PRIMATE RESEARCH

In the final section, the authors discuss public perceptions of primate research and how such perceptions or misperceptions can be corrected. Bennett (chapter 19) argues that scientists must participate in the public arena and gain the confidence of the public, if research is to continue at the present rate. He emphasizes the need for a scientific dialogue with regulators and the importance of stressing the need for additional research in areas where the data are scarce or equivocal.

Perspectives on public response to animal research are provided by representatives for moderate animal welfare groups. Dodds (chapter 20), representing the Scientists' Center for Animal Welfare (SCAW), notes that the research community has made significant strides over the last 10 years in the area of self-regulation. Leonard, a veterinarian for the Animal Rescue League of Boston, describes, in chapter 21, the regulatory process in the state of Massachusetts in which veterinarians of designated animal welfare groups (i.e., the Animal Rescue League of Boston and the Massachusetts Society for the Prevention of Cruelty to Animals) conduct state inspections of research facilities. Leonard further suggests that such an arrangement might be successful in enforcing federal regulations.

In chapter 22, Carroll, Director of iiFAR (The Incurably Ill for Animal Research), characterizes public perception of primate research and provides several suggestions for scientists when they encounter the media or the public.

In the final chapter, King argues that scientists can no longer remain in their ivory towers; rather they have an obligation to educate the public about the benefits of animal research. This educational effort can and should involve national scientific and professional organizations. More importantly, however, much of this effort should be pioneered by universities through existing channels (outreach programs to secondary schools, newsletters to alumni, news releases, and other public relations activities). Failure to make these efforts both institutionally and individually may alter the process and progress of science.

CONCLUSION

The papers collected in this volume represent important explorations of the world through the looking glass. As Alice discovered, her reflection constituted only a limited and in some cases distorted view of that world. In a similar manner, psychological well-being must be considered along a number of different dimensions, not only from the historical and administrative to the philosophical and scientific, but also from our own unique orientation to that of the nonhuman primate's perspective.

Understanding psychological well-being in nonhuman primates is a difficult proposition at best. Writers of the new regulations have assumed that specific structures can readily be identified that will promote psychological well-being in nonhuman primates. However, these assumptions are currently based more on human feelings than on the actual reactions of nonhuman primates. Researchers argue that it is the response of the animals themselves to environmental manipulations that should determine the effectiveness of those manipulations. In this regard, much work needs to be done. The effects of various environmental manipulations on the psychological well-being of many different primate species should be evaluated in scientifically controlled studies.

Some animal-rights activists will argue that researchers are merely stalling and are trying to avoid this new regulatory challenge. On the contrary, researchers are eager to arrive at enforceable regulations, because they have a primary obligation to maintain the highest level of functioning of the nonhuman primates and other animals with which they work. However, the regulations should be effective in achieving the desired goal, not merely cosmetic. To achieve effective regulations, we cannot passively accept our own reflection of the world of nonhuman primates, but must, like Alice, step through the looking glass to obtain the animals' perspective.

References

Animal Welfare Act, 7 U.S.C. §§ 2131-2157 (1985).

U.S. Dept of Agriculture, Animal and Plant Health Inspection Service, Animal Welfare: Standards; Proposed Rule (9 C.F.R., Part 3) [Docket No. 90-040]. Federal Register 55 (158), 33448–33531; Aug. 15, 1990.

PART ONE

HISTORICAL PERSPECTIVES ON PSYCHOLOGICAL WELL-BEING LEGISLATION

CHAPTER 2

ANIMALS IN RESEARCH: PUBLIC POLICY DETERMINANTS

FRANKLIN M. LOEW

Until 1966, no U.S. federal law dealt with the circumstances or conditions under which animals were studied and used in research and teaching. Since what is now the Animal Welfare Act was passed that year, public concern about the use of animals has steadily increased. Today, animal research is one of the most contentious issues facing science and society. The time has now come for scientists to concede that at least some animal laboratories and animal-care practices did need improvement and that the scientific community did need to pay far greater attention to such issues as the review of protocols, the control or prevention of pain and distress, and the justification of necessary animal research. Critics must accept that the scientific community *is* changing, that many fields of science still do, and may for a long time to come, depend on animals for the medical progress and consumer safety that the majority of Americans reportedly want, and that any regulations controlling animal research will carry a financial cost that is ultimately borne by the public purse.

THE ANIMAL WELFARE ACT

The first set of regulations for the first—and still the only—federal law dealing exclusively with animals in research went into effect over twenty years ago, in 1967. The Laboratory Animal Welfare Act (1966), controlling dealings in and acquisition of laboratory dogs and cats, was passed in 1966 after stormy and dramatic hearings in both the House of Representatives and the Senate. The act was passed where earlier initiatives had failed, largely because of a public scandal about the deplorable conditions in which animal dealers kept their dogs and cats, and because the act, itself, was very limited in its intent. The ensuing 20 years have been filled with more hearings. Several amendments to the original law (now known as the Animal Welfare Act, 1970) have broadened and expanded its scope to address care (but not use) of animals in the research facilities and have increased requirements for accountability regarding animal use by the National Institutes of Health and other federal agencies which support research.

This same 20-year period has seen similar changes in laws and more restrictive regulations regarding research using human subjects and research involving recombinant nucleic acids. The "animal issue," however, is different in several ways. For 200 years, a small but impassioned and vocal segment of the public has protested the use of "helpless and innocent" animals in laboratories. In the first half of this century, these voices were represented by the antivivisection societies (supported by William Randolph Hearst's newspapers). However, in the last 20 years, a much broader segment of the public has begun to voice concern about animal research, joined by philosophers, some scientists, and other professionals. Like the abortion issue, the animal-research controversy evokes a depth of passion from both protesters and supporters that is not seen in either the human subjects research or the recombinant nucleic acid debates, probably because the animal-research controversy involves less reliance on facts and more on values.

As a result, animal research has evoked more public and congressional interest than have the other two issues, especially in the last 5 to 10 years, and the policy implications may be even more interesting. Public attitudes toward animals, and even the idea or concept of "animal," have changed markedly in a generation, and a generational division among the scientists themselves has emerged. Public concern appears to have increased as a result of greater acceptance, via television and the popular media, of the notion that animals are cognitive beings with thought processes that, if not the same as those of humans, are at least sufficiently similar to raise questions about animal treatment in laboratories and elsewhere.

Scientists have not been isolated from such societal changes. Broadly speaking, scientists who were trained and had at least some experience in an animal laboratory before the initial 1967 animal-research regulations are all necessarily at least 45 or 50 years old today. These scientists were accustomed to being able to do (to an animal) whatever they deemed necessary to achieve their research goals. Today, however, the methods of research are coming under increasing scrutiny. Thus, younger scientists, especially those starting in the last 5 to 10 years, can be presumed to have known no context for their research except a regulated one. They do not appear to be quite so concerned about the effects of more stringent regulations, and some of them are active supporters of animal-protection groups.

ANIMAL RESEARCH CONTROVERSIES

The history of animal-research controversy in the last dozen years is essentially episodic, punctuated with particular cases producing surges of protest and then reaction. In 1973 and 1974, there was an outpouring of public protest against the use of dogs by the Army and the Air Force. In particular, proposed Army nerve-gas experiments on dogs produced a flood of 3,000 letters to the Pentagon—reportedly more than were received when MacArthur was fired by Truman. This protest was uncoordinated, but two years later, Henry Spira, a New York English teacher, organized and led a concerted attack on cat sexual behavior experiments conducted at the American Museum of Natural History. Spira had done his homework before beginning the action and had also approached the Museum to ask it to stop the experiments. Not unexpectedly, he was rebuffed, but this rebuff only strengthened his hand with the media. After 18 months of campaigning, during which the museum lost donors and was the focus of considerable negative publicity, the research scientist involved in the cat experiments retired, and the museum closed the laboratory.

The decade of the 80s has seen many more focused campaigns. Spira switched his attention to the use by the cosmetic and household-product industries of the Draize eye-irritancy test, in which test substances are placed in the eyes of rabbits and the extent of subsequent inflammation and tissue destruction is evaluated. Gathering a loose coalition of 400 animal-protection organizations, Spira concentrated attention on the Revlon company. After a 12-month campaign, the company announced a $750,000 initiative to seek an alternative to the Draize test. Other companies rapidly followed suit. To date, over $5 million have been spent on the search for an alternative, cosmetic companies have reduced or even eliminated their use of animals, and some promising cell-culture tests have been developed. More importantly, all segments of the chemical, drug, and household-product industries have started to invest in the development of cell-culture technology.

At the time (1980) that Spira was launching his Draize-test campaign, a group called United Action for Animals was promoting a bill in Congress to fund the development and application of alternatives to animal research. The bill stirred the sleeping animal-welfare giant said to lurk among the general public, and an outpouring of support forced Congress to address the topics of alternatives and the way in which research animals are *used* (as opposed to the topic of their care only, which was already covered by the Animal Welfare Act). However, shortly before scheduled hearings took place, a new animal-research scandal erupted. In September 1981, a private laboratory receiving grant support from the National Institutes of Health (NIH), the Institute for Behavioral Research (IBR), was raided by state and local police. Seventeen monkeys and other evidence were removed, and a well-known neuroscientist was charged with animal cruelty under the Maryland anticruelty statute. The charges were ultimately dismissed on a technicality by the state appeals court, but NIH found violations of its own guidelines on the care and use of animals in research and terminated the IBR grant. This incident, more than any other, persuaded some members of Congress that something more had to be done, although there was still considerable opposition to more regulation.

A second, widely publicized incident at the University of Pennsylvania seemed to convince many others that the Maryland case may not have been an isolated infraction. Videotapes of research activities in a head-injury laboratory, made by the investigators themselves, were stolen and versions of them released to the public. The tapes provided convincing visual and graphic evidence of inadequate surgical procedures and a mocking attitude by technicians and students toward brain-damaged baboons. NIH again found violations of its policies. The research in the head-injury laboratory was suspended, and wide-ranging administrative changes were made in the university's animal care and use procedures.

These actions certainly focused public and scientific attention on the issue and effectively put the scientific community on the defensive. Critics of animal research had been previously dismissed as either ill-informed "do-gooders" or as antiscientific misanthropes, but now things had changed. The press, other media, and members of Congress were taking seriously the charges that scientists were mistreating animals used in their studies. And some even took seriously the more recently promoted notion that animals have rights, such as the right not to be eaten or to be experimented upon. This "rights" view contrasts with the older and still more widely accepted "welfare" view that permits the killing of animals and their use to satisfy human ends, provided that animal pain and distress are minimized.

Scientists who attempted to recognize the new sensitivities admitted that the Animal Welfare Act needed improvement. In 1985, both the act and the NIH's requirements for grantholders were amended and made more stringent. The relatively broad freedom that scientists used to have in deciding what they needed to do to animals in their research has been constrained. Research institutions have been told to increase their attention to animal care and to review and approve all animal projects prior to their being started. In addition, much greater attention is being paid to the reduction of animal pain and distress, and there is an informal, but growing tendency to attempt to weigh the costs of the research (animal distress and death) against its anticipated benefits (the advance of biomedical practice or knowledge).

CURRENT ISSUES

Where do things stand now? In my opinion, federal law (the Animal Welfare Act, 1985) and the major life-sciences granting agency (NIH) now have adequate powers and procedures in place to assure Congress, members of the public, and scientists that animal-care-and-use programs are appropriate. However, institutional animal-care-and-use committees are still faced with more questions than answers on such seemingly simple but, in reality, very complex questions as evaluating the extent of animal pain and distress on one hand and the potential benefits of research on the other. Also, enforcement by the U.S. Department of Agriculture of the existing Animal Welfare Act regulations has been spotty. Presidential efforts between 1981 and 1987 to eliminate funding for Animal Welfare Act enforcement have always failed in the face of opposition from an unlikely coalition of medical research and animal-protection organizations.

As with nearly all regulations, there are costs to society related to their enforcement. I wish to address three points not dwelled upon in recent reviews of this subject and to attempt to identify some policy implications or alternatives.

Expense of Regulation

Whether those who proposed them intended so or not, the increased federal (and even in some cases, state) standards for animal research and toughened review procedures for institutional animal-care committees are raising the dollar costs of much agricultural, pharmaceutical, medical, and other life-sciences research. In addition to the obvious implications for the national cost of research itself, it is possible that both university- and industry-based research in agriculture, medicine, biology, and biotechnology may tend to move away from perceived high-cost (i.e., highly regulated) jurisdictions to lower-cost ones (i.e., less regulated).

Internationally, the European Economic Community (EEC) is discussing a new, unified set of standards which, if eventually similar to those of the United States, Canada, and other developed countries, ought to prevent "regulations-dodging" by multinational companies, who can at present follow the less stringent regulations of other members of the Organization for Economic Cooperation and Development (OECD). Japan is a notable exception to the trend among OECD nations regarding the trend toward more regulations on animal research. But what about developing countries desirous of attracting major biotechnology or pharmaceutical companies with their smokeless factories and high-tech work forces? It is possible that the United States will indeed lose some industrial research and development to these countries because of our new animal-research restrictions. On the other hand, biomedical research increasingly requires high-technology support services and skilled technical staff, both of which tend to be scarce in developing nations. As a result, although pharmaceutical manufacturing plants may be located in less-developed countries, research facilities are concentrated in North America, Western Europe, and Japan.

Although OECD nations may have similar levels of regulatory control over animal research, there are still differences in national history and philosophy that may lead to different regulatory approaches. For example, individual scientists in the United Kingdom have to have licenses to perform animal research, under a recently revised law originally passed in 1876. In the United States, it is the institution, not the individual scientist, which has the legal relationship with the government and the granting agencies. Most scientists working in the United Kingdom have come to terms with scientific life under their 1876 Cruelty to Animals Act. Britain certainly compares well with the United States on Nobel prizes in chemistry and physiology or medicine, although some activities (e.g., surgical science or trauma investigations) are more constrained in Britain. Culture is important in science, and British scientists have repeatedly told me that their 1876 act is appropriate to their culture. In 1986, the Animals (Scientific Procedures) Act was passed and is gradually replacing the 1876 act. One of its main points is to add to the need for each scientist to be licensed and the need for each *project* to be licensed as well. In addition, all projects must be subjected to a cost-benefit analysis, and there is some question as to whether cosmetic testing on animals will be permitted.

Public Perception

Scientists and companies whose work requires the study or use of animals need to come to terms with the increasing ambivalence of large segments of the U.S. populations regarding certain, if not all, kinds of animal use. Congress, which appropriates the public funds that are eventually disbursed as grants by the U.S. Public Health Service, the nation's largest funder of life-sciences research, finds this issue to be one of enormous concern to many of its constituents or their organizations. So, while the public and Congress still clearly agree that medical research and its promise of treatments, cures, prophylaxis, and longer life represent a major value in U.S. society, they also seem to demand that less of it come at the nondollar cost of animal pain or discomfort. And nowhere is this demand better shown than in the succession of highly publicized, politically effective events surrounding certain types of experiments on dogs, cats, rabbits, or nonhuman primates since 1980.

American attitudes toward animals have changed in the past decades. The U.S. population is now 98% nonagricultural, and the contact most people have with animals does not extend much beyond pets (there are well over 100 million pet dogs and cats in the United States today) or urban wildlife. As a result, utilitarian attitudes have diminished, and human-animal interaction is increasingly romanticized. In addition, the new science of ethology and increasing attention by television to animal behavior, animal cognition, and animal intelligence have raised concerns about the treatment of animals in modern industrial societies, where even animal husbandry has fallen prey to the methods and metaphors of factories and assembly lines.

Public opinion polls reveal some of the changes in our attitudes to animals and animal research. In 1949, a national opinion poll found that 85% of the public supported the use of dogs in research. By 1983, that figure had fallen to 43%. On the other hand, in 1989, about 29% of the general public in the United States disapproved of *all* animal research, with 64% offering qualified support (''Most people say,'' 1989). In the United Kingdom, however, 40% of the public are opposed to animal research (''Opinion poll,'' 1989). Historically, attitudes to animals in the United States have tended to follow those in Britain, so there is obviously no room for complacency on the animal-research issue. Scientists and their organizations must not only ensure that the new regulations work but that they are seen to be working. Members of the public, including critics of modern research, must have appropriate access to laboratories and a role in the formation and implementation of policy. A knee-jerk response, attempting to exclude critics and defend all animal research, is counter-productive in the face of continuing suspicion of scientific motives and of repeated exposures of cases where researchers either appear to be or are callous regarding the interests of the laboratory animals.

The public also differentiates among different animal species, with rodents eliciting relatively little sympathy, and dogs and primates being highly favored. For 40 years, the research community has been using unclaimed dogs from city pounds and shelters (otherwise scheduled to be put to death) in research, under so-called ''pound seizure'' laws that mandate release of unclaimed animals to registered research institutions as needed. The animal-protection movement has been fighting this practice for just as long. In the 1940s and 1950s, the research community generally was victorious. Now, the animal-protection movement is winning.

By and large, the battle is mainly fought around sentiment and semantics. For example, polls indicate that a majority of the public (about 71%) support the release of shelter animals that are to be killed to research institutions (''Most people say,'' 1989), while at the same time they oppose the release of ''former pets.'' Although many of the 5-6 million dogs killed annually in pounds and shelters are former pets that have been discarded by casual or irresponsible owners, they are also shelter animals that are going to be killed. Regardless of the scientific and other merits of the case (and different arguments can be used to support both scientists and the opponents), the use of pound dogs is such a political minefield that both funds and fortitude are required for any who decide to enter the fray. Primates, or at least chimpanzees, are now in a similar category. Today, the attitude toward the family pet is such that more and more cats and dogs are being bred just for research, to supply a portion of the 200,000 or so dogs and 80,000 cats required annually. Of course, this practice is expensive in money as well as in animal lives.

The Middle Ground

Finally, I hope that an emerging middle ground on these animal-research-related issues will not be destroyed by extremist views or organizations at either end of the spectrum. Most scientists know or can guess the various arguments advanced by critics or defenders of animal use in science, and I will not repeat them here. There are, nevertheless, certain categories of people and positions that need to be sorted and blended (or excluded) to form, if not a map, then at least a direction, according to which 21st-century animal research might proceed:

1. The scientist, probably over the age of 50, to whom the very existence of this issue represents neo-Luddism, and who feels, the wishes of Congress notwithstanding, that it ought never to be taken seriously
2. The younger scientist who, while concerned over the arguments and the sometimes violent tactics of the critics, accepts regulation as a fact of scientific life and develops strategies to cope with restrictions on animal use and care
3. The animal activist or scientist who presses for better animal care, better review (by scientists) of proposed research, and greater public accountability by granting agencies, individual universities, hospitals, companies, and their scientists
4. The animal activist who believes that animal research should halt, either for ethical and moral reasons or for perceived or alleged scientific reasons

Holders of positions (2) and (3) have been coming closer and closer together in recent years. While compromise is difficult, it is surely the better public policy alternative. The facts are that animals are being used less and being given far better care in laboratories today than ever before, and the regulatory agencies are better able—politically, and sometimes financially—to deal with noncomplying scientists or institutions.

At this point in U.S. scientific history, I believe that scientists should seriously address the concerns of American legislators, if not those of voters. No national scientific enterprise today can flourish without the enthusiastic support of the society of which it is a part. And if science is not value-free, then one of the emerging values of U.S. science—a value long prominent in the United Kingdom—is that the animals that have formed such a vital part of many of its accomplishments must be accorded a consideration more concordant with the apparent values of late-20th century America as a whole.

More laws or regulations beyond those already proposed are not needed right now (except, in my view, for the inclusion of mice and rats under the Animal Welfare Act's regulations). Instead, we need to support the recent regulatory changes, implement them carefully, monitor their costs, and encourage compromise rather than confrontation. Listening to the other person is a necessary condition for compromise.

References

Animal Welfare Act, 7 U.S.C. §§ 2131-2157 (1985).

Animal Welfare Act, Pub. L. No. 91-579, §2, 84 Stat. 1560 [Codified as amended beginning at 7 U.S.C. § 2131 (1970)].

Laboratory Animal Welfare Act, Pub. L. No. 89-544, §1, 80 Stat. 350 [Codified as amended beginning at 7 U.S.C. § 2131 (1966)].

Most people say animal research is necessary. (1989, June 9). *AMA News*, p. 2.

Opinion poll results—Cause for concern. (1989, October). *RDS Newsletter*, pp. 1-2.

CHAPTER 3

THE NECESSITY FOR INTERPRETATION OF STANDARDS DESIGNED TO PROMOTE THE PSYCHOLOGICAL WELL-BEING OF NONHUMAN PRIMATES

RONALD D. HUNT

FOCUS ON INTERPRETATION

The original Laboratory Animal Welfare Act (1966) and its recent amendments as enacted by the 99th Congress of the United States contain many unambiguous requirements, which, as laws, do not require interpretation ("Test of the Improved Standards," 1985). In fact, there is no room for interpretation. For example, the amendments enacted in 1985 state that each research facility must have a committee; that the committee must be composed of no fewer than three members; and that the committee must be charged with specific obligations concerning animal care and research. On the other hand, the law states that standards should include minimum requirements for a "physical environment adequate to promote the psychological well-being of primates" (Animal Welfare Act, 1985, Section 13). Neither Congress, which passed the amendments, nor the Secretary of Agriculture, who is charged with promulgating regulations to enforce the act, nor the research community can define "psychological well-being.'' Hence, none of us fully understand what physical environment will promote it (Novak & Suomi, 1988).

This component of the law demands interpretation, as will the relevant standards or regulations. These conference proceedings are a part of the interpretive process, and the papers in the book underscore our lack of understanding of what ''psychological well-being'' is or how to promote it throughout the complex, multispecies order of primates.

The Animal Welfare Act (1985) begins with the following statement:

> the Congress finds that 1) the use of animals is instrumental in certain research and education for advancing knowledge of cures and treatment for diseases and injuries which afflict both humans and animals. (§ 2131)

We cannot lose sight of this finding. We cannot forget or ignore the primary purpose of using laboratory nonhuman primates, nor the obligations and responsibilities of the biomedical research community toward meeting national goals of biomedical research. The standards must be balanced with research protocols. We must recognize that compromises will have to be made so that, in the final outcome, our definition of *psychological well-being* and the appropriate environment for its promotion do not thwart medical progress.

The provision of what is ultimately determined to be an appropriate environment to promote the psychological well-being of nonhuman primates will obviously not ensure the well-being of each species of nonhuman primate or of each individual. The regulations are very unlikely to address the uniqueness of the multiple species in the order Primates and will have to allow for flexibility of interpretation. The individuality of each primate with respect to background, sex, genotype, and so forth must also be considered. It is one thing to assess the status of populations with breeding statistics, longevity, and psychological measurements, but finer interpretation will be required in assessing the psychological well-being of an individual monkey and, of equal importance, the significance of that well-being. Medicine is not an exact science, and mental health is one of the less exact of the disciplines within medicine, especially within veterinary medicine. Interpretation of the well-being of nonhuman primates must draw, in part, from experiences with human mental health, just as primate physical well-being is inferred from experience with human physical health.

What do we know of human mental health? It is estimated that at least 40 million people in the world suffer from severe mental illnesses, such as schizophrenia and severe depression, and at least twice this number are seriously disabled by mental retardation, dementia, and other disorders of the nervous system (Hollzman, Evans, Kennedy, & Iscoe, 1987). Estimates vary as to the number of people in the world affected by the less severe, but nevertheless, incapacitating mental disorders; no estimate, however, is lower than 200 million. Of the 400 million disabled people in the world, no fewer than two out of every five are incapacitated by mental or neurological disease (Hollzman et al., 1987).

The causes of mental disorders in human beings are poorly understood, and the means of rehabilitation or prevention are under constant debate. The mental health of monkeys is an unknown. We are embarking on a relatively unexplored field. Many of the papers from this conference describe behavioral characteristics of various nonhuman primate species, with emphasis on environmental influences. None, however, addresses the influence of organic disease on behavior. The potential causes of organic brain disease are numerous. For example, nonhuman primates contract a great diversity of viral infections such as those caused by cytomegaloviruses (CMV), which in humans are a leading cause of mental retardation. The separation of true behavioral disease from organic disease is extremely difficult and is one of the greatest challenges for psychologists and psychiatrists. The role of organic brain disease in the behavior of nonhuman primates is a topic largely unexplored and largely ignored in behavioral studies.

No matter what regulations may be forthcoming, what becomes important to the interpretive process is that a mechanism should be in place to allow for continued interpretation or assessment, so that the regulations may be continually modified or fine-tuned (or, more aptly, coarsely-tuned) as required. The Department of Agriculture has alluded to this need with respect to underregulation. It is to be hoped that the concept will also exist for overregulation.

HISTORICAL PERSPECTIVE

The original Animal Welfare Act and most of its subsequent amendments drew upon a body of knowledge responsibly developed by the biomedical community. I would like to point out a few historical facts on what has occurred in this century in medical schools and research institutions in the United States, and on how the research community itself has been instrumental in formulating standards of laboratory animal care that only later became incorporated into law.

In 1904, when the National Institutes of Health (NIH) moved from Staten Island in New York and reopened in Washington, DC, its second director, Dr. Milton Rosenau, issued rules on the operation of the laboratory and the conduct of research, which included, and indeed emphasized, the humane treatment of laboratory animals, with particular emphasis on eliminating pain (Furman, 1973). These rules or policies, which were promulgated when animal research was in its infancy, 81 years prior to the current Public Health Service policy on humane care and use of laboratory animals, did not become effective until December 31, 1985.

On November 2, 1907, the Faculty of Medicine of the Harvard Medical School established a Committee on Animals (1911). The charge as it still stands reads:

> Established by the Faculty of Medicine on November 2, 1907: A Committee of at least five, which have power to inspect the animal rooms at any time, to issue any orders and regulations they may deem fit for the proper care and feeding of animals, to discharge any employee of the School who in any way treats animals with cruelty, to dispose of any animal which is in any way objectionable and to determine the sources from which animals are to be obtained. This Committee shall also have general power to stop any experimental work on animals that they may consider improper or unnecessary. (pp. 8–9)

This policy is not altogether different from what is now mandated by Public Health Service policy or by the 1985 amendments to the Animal Welfare Act.

Between 1904 and World War II, medical research, including that employing animals, did not grow rapidly. In fact, the NIH, the major sponsor of research, was not officially established by Congress until May 26, 1930, and the cornerstone of the first building of the present NIH campus was not laid until June 30, 1938. In 1904, when the Director issued his rules on animal care, the NIH was known as the Hygienic Laboratory.

It was not until after the Second World War that intramural and extramural programs of the NIH expanded, and medical research entered a new era that rapidly grew to the present-day level. Society demanded and government responded with an increased national effort to combat disease. During this phase of growth, the medical research community recognized the need for specially trained personnel to manage animal care, and throughout the 1950s, medical schools began to organize units, divisions, centers, or departments of laboratory animal medicine. This reorganization came about as a result of their own initiative, their own recognition of need; not as the result of social pressure or Public Health Service mandates or Congressional legislation.

Those veterinarians who accepted the challenge presented to them by the research community pioneered a new specialty, which led to the establishment, in 1957, of the American College of Laboratory Animal Medicine. This college constituted the third specialty board recognized by the American Veterinary Medical Association. It was established long before the Colleges of Veterinary Internal Medicine, Veterinary Surgery, Veterinary Radiology, and others.

In 1963, the *Guide for the Care and Use of Laboratory Animals* was prepared by the Institute of Laboratory Animals of the National Research Council of the National Academy of Sciences under contract from the Animal Resources Program of the Division of Research Resources of the NIH and published by the Department of Health and Human Services (National Research Council, 1985). The contents of the Guide, defining standards of appropriate laboratory animal care, were largely gathered from observations and research by members of the new veterinary specialty working predominantly at medical schools. This publication, most recently revised in 1985, has provided the basis for almost all existing rules and regulations.

In 1965, the scientific community established the American Association for Accreditation of Laboratory Animal Care (AALAC) to provide a mechanism for peer evaluation of animal-care programs. It is a voluntary program that has worked well. By the late 1960s, each of the major medical schools in the United States had within their organizational structure a program in laboratory animal care to include a veterinarian on the faculty; a committee on animals; and guidelines and rules on animal care and research. They were also becoming accredited by AALAC. Today almost 500 institutions are fully accredited.

These events, evolving from the early 1900s and continuing today, have provided the basis for understanding priorities in laboratory animal care and, as previously stated, have provided the basis for almost all existing laws, regulations, and policies. Clearly, biomedical research institutions have led the way and have done so responsibly.

In 1966, Congress passed the Laboratory Animal Welfare Act (1966)—the first legislation specifically addressing laboratory-animal care. Those portions that apply to research institutions are not dissimilar from the NIH Guide. Congress merely took what the research community had developed and made it applicable to all institutions.

There is no published guide for the ''psychological well-being of nonhuman primates,'' a lack that probably explains why the 1985 congressionally enacted amendments are so unbalanced. Very specific and detailed provisions are included in the law with respect to committee structure and function, various aspects of veterinary care, training, and so forth, but psychological well-being is handled with a clause of about a dozen words.

It is unfortunate that the research community had not prepared an appropriate guide for Congress and the Department of Agriculture to follow as it had for other parameters of laboratory animal care. However, the guide for psychological well-being is now being written. It will ensue from the proceedings of this and subsequent conferences. We know from the papers in this volume and from journal articles we have read over the past several years that the biomedical community has not ignored the subject. A wealth of knowledge has been accumulated, and procedures had been implemented to varying degrees long before the stroke of the Congressional pen.

Now it is important to interpret appropriately what we know, to continue research to expand our knowledge base, and to provide a mechanism for continued interpretation and modification of regulations.

References

Animal Welfare Act, 7 U.S.C. §§ 2131–2157 (1985).

Furman, B. (1973). *A profile of the United States Public Health Service 1798–1948* (DHEW Publication No. 73–369). Washington, DC: U.S. Department of Health, Education, and Welfare, National Institutes of Health, National Library of Medicine.

Harvard University School of Medicine. (1911). *Minutes and records of the faculty of medicine, 7*, 8–9.

Hollzman, W. H., Evans, R. I., Kennedy, S., & Iscoe, I. (1987). Psychology and health: contributions of psychology to the improvements of health and health care. *Bulletin of the World Health Organization, 65*, 913–935.

Laboratory Animal Welfare Act, Pub. L. No. 89-544, §1, 80 Stat. 350. [Codified as amended beginning at 7 U.S.C. § 2131 (1966)].

National Research Council. (1985). *Guide for the care and use of laboratory animals* (NIH Pub. No. 85–23). Washington, DC: U.S. Department of Health and Human Services.

Novak, M. A., & Suomi, S. J. (1988). Psychological well-being of primates in captivity. *American Psychologist, 43*, 765–773.

Test of the Improved Standards for Laboratory Animals Act. (1985, December 7). *Congressional Record.*

CHAPTER 4

THE 1985 ANIMAL WELFARE ACT AMENDMENTS

DALE SCHWINDAMAN

These are indeed crucial times for medical researchers in general and for primate researchers in particular. Animal-welfare activism and animal welfare legislation are affecting this work as never before. In fact, a prestigious journal noted last year that we are in the midst of an era of animal activism unprecedented in this century. I agree that activism on behalf of animals is at its highest level yet. I also believe this activism is the natural result of the changing role of science within society and the great influence that science and society have on each other. In such times, it is important that we in the federal sector should meet often to exchange views with researchers and their colleagues.

Social values determine the nature of scientific inquiry—and the results of that inquiry not only improve the quality of life, but also alter our very lifestyle and the ways in which we understand the natural world and our humanity within that world. Certainly a major contribution of scientific inquiry to society has been an increased awareness and respect for biological life in all its dimensions. Science has given us an understanding of the delicate balance among life forms, even within a single cell. From this increased awareness, ethical values have been shaped, including a renewed respect for the fragility of biological and, certainly in recent years, psychological existence. The treatment of experimental subjects, as well as the process of investigation itself, therefore becomes subject to such values.

If you doubt the evolution of ethical values, contrast Jenner's use of human subjects in testing the first vaccination for smallpox with today's methods. Legend holds that Jenner performed his first vaccination on a passing youngster—a lad simply picked up off the streets. Fortunately, the past 200 years have brought enormous changes in our views about medical experimentation and an increased sensitivity toward human subjects. The watchword now for scientists using human subjects is voluntary, informed consent. It is not surprising, then, that with increased sensitivity for human rights in medical research has also come a concern for the welfare of animals used in such research.

This sensitivity is continually being sharpened by the exposure given to animals by today's visual media. Television specials thrill us with our first look at the birth of a panda; a *Bambi* movie charms and delights us; and a television news report on neglected or abused animals—whether they are being exhibited, transported, or used for research—fills us with anger and frustration.

APHIS AND THE ANIMAL WELFARE ACT

The agency I represent, the Animal and Plant Health Inspection Service (APHIS) of the U.S. Department of Agriculture (USDA), is charged with enforcing the Laboratory Animal Welfare Act (1966), the federal law that emerged from this value system—call it animal-abuse awareness, if you will. Section 1751 of the 1985 amendments to that law clearly reflects society's sensitivity to the use and care of animals in biomedical research:

> Subtitle F—Animal Welfare
>
> Section 1751. For the purposes of this subtitle, the Congress finds that—
>
> (1) the use of animals is instrumental in certain research and education for advancing knowledge of cures and treatment for disease and injuries which afflict both humans and animals;
>
> (2) methods of testing that do not use animals are being and continue to be developed which are faster, less expensive, and more accurate than traditional animal experiments for some purposes and further opportunities exist for the development of these methods of testing;
>
> (3) measures which eliminate or minimize the unnecessary duplication of experiments on animals can result in more productive use of Federal funds; and
>
> (4) measures which help meet the public concern for laboratory animal care and treatment are important in assuring that research will continue to progress. (Food Security Act, 1985)

Just as these concepts grew out of society's concern for research animals, so did the original Laboratory Animal Welfare Act of 1966. That act was aimed at ending the practice of "petnapping," and it regulated interstate traffic of stolen animals when they were used for research.

A far more comprehensive Animal Welfare Act was passed four years later in 1970 as a series of amendments to the original 1966 act. It expanded coverage to include most warm-blooded animals used in research, zoos, and circuses; marine animals used in shows and exhibits; and animals sold in the wholesale pet trade. Further amendments to the act in 1976 extended its scope to include the care and treatment of animals while they are being transported. The 1985 amendments once again focused on research animals.

The secretary of the USDA was appointed by Congress to promulgate the Animal Welfare Act and its amendments. The secretary has delegated the enforcement to APHIS, which enforces the act through a system of licensing, registration, and unannounced inspections of regulated businesses—those businesses that buy or sell warm-blooded animals, exhibit them to the public, transport them commercially, or use them in research. Persons who violate the act's standards of care and treatment are charged and prosecuted through either civil or criminal procedures.

To give you some idea of the scope of our work, during fiscal year 1987, APHIS inspectors checked 6,732 facilities an average of two times per year. In the area of medical research, APHIS received reports from 1,260 nonfederal and 160 federal research facilities using laboratory animals covered by the act. The reports showed that of all the laboratory animals protected under the Animal Welfare Act, 68.2 percent were not exposed to painful or distressing procedures, 25.2 percent were administered anesthetics, analgesics, or tranquilizers, and 6.6 percent were not given pain relief because to do so would have interfered with test results. To carry out its work, APHIS was given a budget of $6.2 million in 1988 as compared to $5.9 million for fiscal year 1987. Both figures include $750,000 to operate the USDA's Animal Welfare Information Center (AWIC).

THE 1985 ANIMAL WELFARE ACT PROVISIONS

The 1985 amendments (Animal Welfare Act, 1985) have set out new and rather stringent conditions that focus on research animals. There are four key areas of the amendments requiring:

- Adequate exercise for dogs as determined by an attending veterinarian, and the establishment of a written program and system for carrying out this requirement;
- A physical environment promoting the psychological well-being of nonhuman primates and a written program and system for carrying out the requirements;
- Specific methods for animal care, treatment, and practices in experimental procedures to ensure that animal pain and distress are minimized. This involves adequate veterinary care, the appropriate use of pain-relieving drugs, the consideration of alternatives to painful procedures, and a limitation on the use of animals in survival surgery; and

- An institutional animal care and use committee (IACUC) at every research facility, with a number of additional requirements concerning the membership and operation of this committee.

In March of 1987, after getting input from all interested parties, APHIS published proposed rules for all amendments except those affecting dogs and primates. We received almost 8,000 comments—more than APHIS has ever received on a proposed regulation. After careful review of all the comments, we drafted our final rule on these amendments, together with a proposed rule on dogs and primates. For the proposed rule on primates, we had valuable assistance from both the American Association of Zoological Parks and Aquariums and a panel of primate experts selected from a list supplied by the National Institutes of Health (NIH). The Agency is satisfied that the new rules will adequately implement the new amendments.

It is important to keep in mind that the intent of the improved standards for primates is to provide adequate space, equipped with devices for exercise consistent with the natural instincts and habits of specific primates. Therefore, requirements may vary depending on the species of primate in question.

Meanwhile, I am aware that the research community is greatly concerned about the increased costs the new regulations will bring. Many are understandably distressed at what seems to be a serious drain on badly needed research funds. APHIS certainly empathizes, for we, too, are faced with escalating costs. APHIS is proceeding on the assumption that the regulations will constitute a major rule; that is, we believe the costs to industry will be in excess of $100 million. And this level of cost, by Presidential order, requires APHIS to undertake a full economic assessment.

THE ANIMAL WELFARE INFORMATION CENTER

I also want to comment on the fact that some members of the public have argued with the congressional findings, calling for more support for developing alternate testing methods and ways to end unnecessary duplication of experiments. They say that if all data were made widely available to the research community, unintended duplication of animal experiments would be avoided.

The dissemination of such data is one of the purposes of the USDA's Animal Welfare Information Center. Still, we at APHIS believe that if principles of sound investigation are to continue, animal contribution to research cannot be discounted. Neither do we believe, as some predict (perhaps out of wishful thinking) that there will come a time when research on animals will be entirely eliminated.

Such controversies are a dramatic reminder to us of the role education must play if society's demands on science are to have positive and progressive results. In this respect, Congress made a logical choice in selecting APHIS, with its many experienced veterinarians and animal health technicians, to administer the Animal Welfare Act. Let me emphasize that the effective management of our animal welfare responsibilities is second to no other APHIS activity. But we must discharge more than just our technical responsibility for administering the law—we must also pursue that necessary effort to educate and communicate.

The most important concept we can offer our clients is the need for good animal husbandry. Proper nourishment, proper attendant care, protection from abuse—to stress all these is to stress the bond between people and their animals. We care about animals—surely that is why we joined our profession in the first place!

The proper care of experimental animals is a sound research principle. It is only common sense, for instance, that an animal will not respond normally if it is stressed or undernourished. Stress can jeopardize the proper metabolism of drugs in the system. Nutrition affects blood chemistry, which in turn affects the analysis and evaluation of testing results. Because the validity of research results is so dependent on the health of research animals, the future of science depends on the integrity of scientific methods and the treatment of animals used in research.

These examples show the very sound logic behind the law, and so the stewardship that APHIS seeks to promote can best be realized by educating people about this logic. In this way, we will *lead* people into compliance instead of forcing them into compliance. And doing so will certainly make our job of administering the law easier!

Teaching sound animal health practices must include the teaching of precise scientific techniques. Just as scientific research depends on sophisticated computer modeling and deft manipulation of complex microscopes, so it also requires adept handling of experimental animals to obtain the samples or tissue needed for accurate data. For example, to reduce the stress and pain of laboratory animals, nontraumatic restraining techniques must be taught. Veterinary knowledge of animal behavior helps create concepts through which appropriate animal restraint is understood. Or, as another example, proper phlebotomy techniques must be taught to avoid the rupture of erythrocytes that would render samples useless.

We believe the teaching of procedural skills is crucial for maintaining high research standards within the laboratory, for ensuring that the law be carried out, and for putting into practice the knowledge we have gained about the natural world through scientific discovery. To help us communicate sound skills and principles, we are cooperating with the American Veterinary Medical Association (AVMA) on a proposed program of certifying or registering a cadre of attending veterinarians as members of a team that will provide education to those working with laboratory animals.

We are also developing a comprehensive training and reference manual containing data on all applicable animal welfare regulation; the principles of humane animal care and handling; information on animal medicine and sciences;

and listings of individuals, organizations, and reference materials to consult for assistance. We will need help in keeping this publication both accurate and current and will be consulting with outside specialists in laboratory animal medicine and related disciplines. Once completed, the manual should be of use to veterinary students, to attending veterinarians at research institutions and, of course, to APHIS inspectors.

THE APHIS MISSION AND PHILOSOPHY

All this, of course, is a reflection of what APHIS is all about—we are more than just a regulatory agency. We are *service*-related, and education is a key ingredient of that service. Moreover, since education is always a two-way street, we cannot render effective service, especially in these times of controversy, without constant feedback from those affected by our regulations. We want to work closely with the research community; we must have your expertise and your experience. After all, to be worthwhile, the Animal Welfare Act must be seen as a living, breathing instrument that can adapt to new information. If something comes along that demonstrates that we are underregulating in a certain area, we need to correct that ommission. If portions of the law are less than clear, concise, and logical, we need to correct that fault as well.

In the final analysis, if APHIS is to honor its commitment to effectively enforce animal welfare legislation, our operations can never be static. We need to make certain that every facet of our agency is organized to meet that challenge. To that end, we are undergoing a major restructuring that is designed to give animal care much greater visibility.

Along with ten other areas of responsibility, our new Regulatory Enforcement and Animal Care staff will report directly to our Administrator, Dr. James Glosser. Instead of falling under our Veterinary Services staff, where it was a small part of an organization with a different mission, the office will now be in a position to compete for its own resources and have a staff of highly trained professionals working full time.

Finally, let me reiterate the APHIS philosophy on animal use. We are committed to minimizing the pain, suffering, stress, and anxiety of animals, especially when such discomfort is knowingly inflicted by people. We want to assure that all the latest drugs, food, care, and procedures that can make life more tolerable for animals, while they are being used for the ultimate benefit to humankind, are being applied. We also support the use of animals for whatever benefit they provide to society, whether that be for food, pleasure, or research.

This is nothing more than the same balanced view reflected in our veterinary oath. While we pledge to take responsibility for the welfare of animals, we also vow to use scientific knowledge and skills for the advancement of medical knowledge. The wise composer of this oath saw no conflict between relieving animal suffering and advancing science. Indeed, there is none.

References

Animal Welfare Act, 7 U.S.C. §§ 2131-2157 (1985).

Animal Welfare Act, Pub. L. No. 94-279, 90 Stat. 417 [Codified as amended at 7 U.S.C. §§ 2131-2156 (1976)].

Animal Welfare Act, Pub. L. No. 91-579, §2, 84 Stat. 1560 [Codified as amended beginning at 7 U.S.C. § 2131 (1970)].

Laboratory Animal Welfare Act, Pub. L. No. 89-544, §1, 80 Stat. 350 [Codified as amended beginning at 7 U.S.C. § 2131 (1966)].

U.S. Dept of Agriculture, Animal and Plant Health Inspection Service, Proposed Rule (9 C.F.R. 1-3) [Docket No. 88-103]. Federal Register 54 (49), 10822–10954; March 15, 1989.

U.S. Dept of Agriculture, Animal and Plant Health Inspection Service, Definition of Terms and Regulations, Final Rule (9 C.F.R. 1;2) [Docket Nos. 89-130; 89-131]. Federal Register 54 (168), 36122–36163; Aug. 31, 1989.

U.S. Dept of Agriculture, Animal and Plant Health Inspection Service, Animal Welfare; Standards; Proposed Rule (9 C.F.R., Part 3) [Docket No. 90-040]. Federal Register 55 (158), 33448–33531; Aug. 15, 1990.

PART TWO

DEFINING PSYCHOLOGICAL WELL-BEING

CHAPTER 5

THE HUMAN MODEL OF PSYCHOLOGICAL WELL-BEING IN PRIMATES

GENE P. SACKETT

"The Apes, vilest of beasts, how like to us"
(Cicero, *De Natura Deourum*).

Well-being and life satisfaction are topics of historical and contemporary interest in the fields of human mental health, social and personality psychology, and mental retardation. Work in these areas serves as a human model for current issues concerning the welfare of captive primates. This essay reviews selected aspects of the human model, outlining some of its implications for research on primate psychological well-being and on the implementation of animal welfare regulations in colonies of captive primates.

Psychological well-being is an idea and an ideal affecting all human lives. The concept is a major theme in religion, literature, philosophy, and social science. Well-being through prevention of conditions producing ill-being has even become a goal in much of modern medicine (Seeman, 1989). The purpose of this essay is to explore some dimensions believed to influence human well-being in order to develop a working model for the study of psychological well-being in captive nonhuman primates. This approach follows that of Harry Harlow (Harlow &

Preparation of this paper was supported by NIH grants RR00166 from the Animal Resources Branch and HD02274 from the NICHHD Mental Retardation Branch.

Mears, 1979), who believed that there was much to learn about monkeys and apes from the study of *Homo sapiens*.

COMPARATIVE MODELS

Comparative models of physical or psychological health have two main goals (Sackett, 1988). The first is to aid in identifying causal processes underlying the target health issue in the species of primary concern. The second is to assist in specifying preventative measures or therapies that will produce beneficial outcomes for the species of primary concern. With respect to human diseases, these goals are often achieved by studying animal models. Such studies begin by describing the known human phenomena of concern. This step is followed by choosing an appropriate animal species that exhibits one or more important characteristics of the human condition or by testing potential model species for the appearance of such characteristics. At this point the model already may be useful for testing therapeutic agents such as drugs or surgical procedures. However, if disease etiologies or mechanisms are at issue, it may be necessary to study causal processes and to assess whether the animal results can be generalized to humans before any practical applications can be made to clinical problems.

Unfortunately, it is not possible simply to reverse this process in the search for a human model of nonhuman primate psychological well-being. Owing to the lack of a shared culture and a system of communication with most individual primates, we cannot even perform the initial model-building step of describing the phenomena in the target species. We can only assume that the indices of well-being expressed by humans also apply to nonhuman individuals. Although this is a dangerous scientific assumption, research on nonhuman primate ill-being suggests that it may have validity. Like humans, monkeys and apes reared in social isolation develop abnormal social, emotional, and communication behaviors (Capitanio, 1986). Separation from mother or peers produces a syndrome of agitation and subsequent depressed behavior in some primate infants that is also seen under similar conditions in humans (Mineka & Suomi, 1978). Thus, the examination of human well-being models may be more than an academic exercise.

Theories of human well-being address three general questions that are also relevant to primates: What aspects of an individual's life should be examined in order to find psychological health, and must they be indexed subjectively or can they be judged accurately by an observer? How, why, and when do individuals experience life positively? What is the relation between perception of one's physical and psychological world and one's well-being?

THEORIES AND DEFINITIONS OF PSYCHOLOGICAL WELL-BEING

Literature and philosophy offer a number of views concerning sources of human happiness. A few quotations from these sources capture the spirit of many current psychological theories concerning human well-being (Diener, 1984). My own favorite is given first.

> *Hedonism.* There is nothing contrived by man by which so much happiness is produced as by a good tavern. (J. Boswell)
> *Social support.* True happiness consists not in the multitude of friends, but in the worth and choice. (B. Jonson)
> *Primary drives.* The foes of human happiness are pain and boredom (A. Schopenhauer)
> *Physical well-being.* A sound mind in a sound body, is a short but full description of a happy state in this world. (J. Locke)
> *Arousal and stimulus change.* A lifetime of happiness! No man on earth could bear it; it would be hell on earth. (G.B. Shaw)
> *Internal subjective state.* Happiness depends, as nature shows, less on exterior things than we suppose. (W. Cowper)
> *Perception and adaptation level.* But, O! How bitter a thing it is to look into happiness through another man's eyes! (W. Shakespeare)
> *Existentialism.* If you want to be happy, be. (L. Tolstoi)

Many of these perspectives raise a critical question for primate well-being issues. Namely, what is the relation between psychological ill-being and psychological well-being? Are these end points on a unidimensional continuum, or do they represent independent dimensions? In his classical study of life-span adaptation in college graduates, Valliant (1977) concluded that "psychopathology is always with us and [well-being] is a way of reacting to problems, not an absence of them" (p. 3). As a general model, this implies that well-being derives from successfully coping with life's problems, not from a situation of having no problems that demand coping behavior. As a model for primates, the statement suggests that a varied and challenging environment should support well-being better than a benign and sterile, though physically healthful, existence.

As suggested by both scientific theory and philosophy, psychological well-being is a subjective phenomenon related to the quality of one's life experiences. Thus, *happiness*, *satisfaction*, *positive affect and morale*, *achievement*, and *success* are some terms used by people to describe the quality of "feeling good." In these terms, psychological well-being can be viewed as both an affective state and a cognitive state.

In his extensive review of research on subjective well-being, Diener (1984) identified three ways in which the well-being dimension of happiness has been defined operationally. Happiness as an emotional experience is often defined as a preponderance of positive over negative affect. This definition involves both satisfaction with life and a typically positive affective state. It thus concerns a chronic rather than acute condition, which could exist even in the face of temporary periods of negative affect. The measurement of happiness under this definition concerns a person's global assessment of his or her general reaction toward life based on subjective internal criteria.

Happiness defined by external criteria deals with the possession of desirable qualities, such as piety or charity, as viewed by an outside observer. In such definitions, a person's behavior or physical appearance is judged against some standard relative to a particular culture or belief system. It should be obvious that there may be little or no correspondence between such external criteria and the affective experiences of the individual being judged.

A third definition of well-being concerns life-satisfaction judgments based on a person's own standards concerning the internal and external circumstances that constitute a satisfactory life. The large amount of research using this definition relies on a person's global assessment of the overall quality of his or her life.

Attributing psychological ill-being or well-being to an animal involves Diener's second definition. The observer views the overt behavior of the animal and, on the basis of his own experience and beliefs, makes a judgment about the animal's internal state. This process is based on the assumption that there is a reasonable correspondence between (a) an animal's overt behaviors and its internal states, and (b) an animal's internal states and the internal states of the human observer. Although neither of these assumptions is actually testable, they both rest on a belief that a human's overt behavior is correlated with that human's subjective state of well-being or ill-being. To the extent that this is true, the human model could apply to primates. Research findings on correlates of human subjective well-being bear on this critical point.

RESEARCH FINDINGS ON HUMAN WELL-BEING AS A MODEL FOR PRIMATES

Diener (1984) reviewed a number of studies on correlates of subjective well-being. It should be kept in mind that the direction of causality in this research is uncertain, as the degree of well-being could affect each of the variables measured.

Biological influences indexed by actual disease incidence had significant but low correlations with subjective judgments of well-being. Subjective judgments of health were more strongly correlated with well-being. This suggests that health indices should not be used as the sole, or perhaps even a major, indicator of psychological ill-being or well-being in primates.

Positive affect, represented subjectively as feeling good, is an important component of well-being. Overall, studies suggest that the frequencies of positive and negative affect episodes are inversely correlated, while the intensities of these episodes are positively correlated. However, in terms of judgments concerning long-term affect, there seems to be little or no correlation between chronic states of positive and negative affect. This finding suggests that indices of acute affective episodes, usually indexed in primates by cardiovascular or hormone measures, may have little or no correspondence to any long-term affective states that may be chronically experienced by primates.

Income is a major correlate of subjective well-being in every country studied. Poor people uniformly report well-being judgments of the same low magnitude, regardless of the absolute level of poverty and life resources in their country. This finding suggests that a discrepancy mechanism involving the differences between a person's economic aspirations and his or her actual attainments is a major determinant of human psychological well-being. It is not clear what this implies for captive primates. Primates will work and learn simply to engage in problem-solving behaviors, with or without external rewards. Perhaps the opportunity to engage in, and succeed at, problem solving could affect a primate's well-being much as income affects well-being in humans.

Demographic factors such as age and sex appear to have little or no correlation with human subjective well-being. Employment and education level are correlated with well-being, but the effects are largely due to the higher income of employed and better-educated people. Although marriage has a modest positive relation to well-being, having children shows either no correlation or a negative relation to well-being. As a model, these results suggest that psychological well-being should be an issue for primate females and males of any age. Close heterosexual relationships are almost certainly a species-dependent issue, and thus a moot point here. However, the idea that adult female primates need to bear and rear offspring to maximize their psychological well-being is not supported by the human model.

Social contact is modestly correlated with well-being, independent of many other factors such as health and socioeconomic status. However, factors such as personality, type of social situation, satisfaction with the individuals being contacted, and prior social experience influence the degree of correlation. Clearly, the human model supports the idea that social contact enhances the psychological well-being of captive primates. However, the human model suggests that not all social contacts, for any groups of animals, at any given time will benefit well-being. Rather, care must be taken to group animals that are compatible to avoid lowering the well-being of some individuals.

Personality factors have shown fairly strong correlations with human well-being. Extroversion and high self-esteem correlate positively with well-being, while neuroticism has a negative correlation. Although self-esteem and neuroticism are difficult to identify in primate behavior, extroverted social and exploratory behavior can easily be measured. Likewise, depressive and withdrawn primate behaviors, behaviors similar to those of some people classified as neurotic, are displayed and measurable. The human model suggests that animals that generally display an extroverted, approach-oriented behavioral disposition should have higher levels of psychological well-being than those displaying a depressed, withdrawn orientation. This model also suggests that episodic and chronic measures of these behavioral dispositions should provide a major index of acute and long-term well-being in primates.

THE SOCIAL PSYCHOLOGY OF HUMAN WELL-BEING

Taylor and Brown (1988) reviewed a number of social psychological studies of human perceptions of well-being. Their review tested the fundamental clinical, mental-health assumption that behaving in accordance with the realities of the physical and social world defines good mental health and psychological well-being, while failure to behave and think in accordance with these realities defines abnormal health and ill-being. They reviewed studies that measured people's judgments comparing their own desirable cognitive and social abilities and personality traits with the same attributes of others. The results showed that subjects who were clinically normal (a) consistently rated their own attributes as better than those of the average person, and (b) rated themselves better than they were rated by others. This finding suggests that most normal people have an unrealistic, illusory view of themselves. Subjects who were classified as clinically depressed or very low in self-esteem judged their abilities to be more average, and their self-ratings were closer to the ratings given them by others. Thus, the self-perceptions of clinical groups exhibiting poor well-being were less illusory and more realistic than those of clinically normal subjects.

These studies have important implications for any model of primate well-being. As discussed above, imputing psychological well-being to animals hinges on the basic assumption that humans and animals share basic subjective processes and their behavioral expression. These social psychology studies suggest that normal humans view themselves and their lives to be better than they are. This normal behavior process presumably serves as a defense mechanism, preserving psychological well-being in the face of real adversity. As a model, this human process suggests that a modest degree of adversity in the lives of healthy primates should not have detrimental effects on their well-being. This probability leads to the prediction that indices of primate well-being should not be negatively affected by a number of physical or social events that appear to humans to be aversive, such as temporary confinement in a small cage or periodic removal from a social group. At any rate, these social psychology studies underscore the complexity of measuring human well-being and identifying its correlates. Certainly, the study of well-being in primates will be at least equally complex.

DEINSTITUTIONALIZATION AND THE QUALITY OF LIFE

The quality of life is a major dimension in the psychological well-being of human and nonhuman primates. A movement to remove people from institutions, begun over 20 years ago, was aimed at bettering the quality of life for persons with mental retardation (Bruininks, Meyers, Sigford, & Lakin, 1981). A guiding principle in this movement is that of the least restrictive environment. The idea is to place retarded individuals into environments, including school and training situations, that maximize the expression of their individual abilities while minimizing intervention and interference from others.

This deinstitutionalization movement has much in common with the psychological well-being movement in captive primates. Both were mandated by federal laws after intense lobbying by activist groups. Both have had large economic consequences. Both have had major social effects on their constituents and on the caretakers of the constituents. Both have forced major changes in housing and other physical facilities. Both have a great need for methods of measuring subjective parameters to assess the effectiveness of these interventions and to develop improved outcomes. Landesman's (1986) discussion of problems in establishing evaluation standards for assessing life quality and life satisfaction in persons with retardation is relevant to the same issues in studying primates.

Landesman and Butterfield (1987) present an analysis of outcomes of different deinstitutionalization efforts that has important implications for the primate psychological well-being movement. They studied a variety of objectively measured criteria involving factors such as social acceptance, return to institutions, and obtaining jobs. Some programs were found to be exceptionally successful, and others were outright failures. Two major dimensions distinguishing successful from unsuccessful programs involved structural factors versus behavioral functions.

Structural factors included administrative procedures, rules governing the behavior of caregivers and residents, and the physical features of buildings. Functional factors involved identification and measurement of desired behaviors indicating competence, adaptive responding, and appropriate affect and emotion, coupled with measures of undesired behaviors involving deviant responding. Programs that were unsuccessful began by designing and implementing structural factors, assuming that the desired behavioral functions would follow. Successful programs began by defining the desired functions, then testing various structural factors for their ability to support those functions.

The primary approach to psychological well-being in nonhuman primates has followed the unsuccessful structural path. Regulations have been promulgated requiring often unattainable administrative procedures, rules have been made requiring caretakers to engage in behaviors that ignore well-known behavioral differences between primate species, and arbitrary cage sizes have been required that may have little to do with the preferences and behaviors of the animals they house. If the human deinstitutionalization model is valid for captive primates, we can expect these structurally based primate well-being efforts to fail also. The message is that goals for desired, measurable behavior must be defined and tested before structural factors are set in regulatory concrete. One path toward learning how to do this might be for primatologists and government regulators to read the existing and future literature on human psychological well-being and deinstitutionalization.

References

Bruininks, R. H., Meyers, C. E., Sigford, B. B., & Lakin, K. C. (Eds). (1981). *Deinstitutionalization and community adjustment of mentally retarded people* (Monograph No. 4). Washington, DC: American Association on Mental Deficiency.

Capitanio, J. P. (1986). Behavioral pathology. In G. Mitchell & J. Erwin, (Eds.), *Primate Biology* (Vol. 2, Part A, Behavior, Conservation, and Ecology, pp. 411–454). New York: Alan Liss.

Diener, E. (1984). Subjective well-being. *Psychological Bulletin*, *95*, 542–575.

Harlow, H. F., & Mears, C. (1979). *The human model: Primate perspectives*. New York: Wiley & Sons.

Landesman, S. (1986). Quality of life and personal life satisfaction: Definition and measurement issues. *Mental Retardation*, *24*, 141–143.

Landesman, S., & Butterfield, E. C. (1987). Normalization and deinstitutionalization of mentally retarded individuals. Controversy and facts. *American Psychologist*, *42*, 809–817.

Mineka, S., & Suomi, S.J. (1978). Social separation in monkeys. *Psychological Bulletin*, *85*, 1376–1400.

Sackett, G. P. (1988). Animal rights, human rights, scientific rights: Who's right? *Contemporary Psychology*, *33*, 23–25.

Seeman, J. (1989). Toward a model of positive health. *American Psychologist*, *44*, 1099–1109.

Taylor, S. E., & Brown, J. D. (1988). Illusion and well-being: A social psychological perspective on mental health. *American Psychologist*, *103*, 193–210.

Valliant, G. E. (1977). *Adaptation to life*. Boston: Little, Brown.

CHAPTER 6

SUBJECTIVE AND OBJECTIVE FACTORS IN ASSESSING PSYCHOLOGICAL WELL-BEING IN NONHUMAN PRIMATES

LEONARD A. ROSENBLUM

If, as a society, we wish to find means of enriching the lives of captive primates, we must, from the outset, reject the notion that there are two opposing camps: one group concerned with "animal welfare" and another different, and perhaps indifferent, group, known as "animal researchers." Those who spend their lives studying primates in various settings must serve as a primary source of the information essential to any effort to understand those features of the environment most conducive to primate welfare. Unsubstantiated accusations to the contrary notwithstanding, concern for the well-being of research subjects generally exists throughout the community of scholars and scientists working with animals. All researchers have, without doubt, been further sensitized by the thoughtful challenges of individuals both within and outside the scientific community, who have stressed our need to be vigilant in attending to the ways in which the living conditions and the treatment of our subjects can be improved. The overwhelming majority of scientists studying nonhuman primates, within either behavioral or physiological spheres, have traditionally been as concerned

This work was supported by USPHS grants MH# 15965 and MH# 42545 and funds from the State University of New York. The author wishes to thank Dr. Michael Andrews and Dr. Ronald Nadler for their thoughtful discussions of many of the issues considered in this paper.

with the welfare of their subjects as any self-appointed guardians of the presumptive rights of these subjects.

As much as anyone, primatologists are interested in and committed to animal welfare—we are at the present time, as we have been in the past, invested in the enhancement of the well-being of our primate subjects. Our position in working with these animals reflects our genuine emotional concern for them, as well as our appreciation of the fact that their contribution to the welfare of the human species ultimately depends upon our maintaining them in a healthful physical and psychological state. The basic approach of the primate research community, in fact, rather consistently reflects a joining of these two complementary attitudes. In this sense, we are all members of the "animal welfare" community.

The differences in perspective that emerge among those concerned with the welfare of nonhuman primates spring from two sources: (a) the basis upon which welfare is to be judged; and (b) the importance attributed to particular welfare considerations in comparison to the perceived benefits for humankind when research requires some compromise with ideal welfare conditions. This second point would require exegesis of an area of moral and humanistic philosophy, clouded by many of the same issues to be discussed below, which although central to ultimate decisionmaking in this area, goes beyond the scope of this paper. The following discussion is directed toward the first point. It is my position that if we are to create the types of environment that serve both the essential informational requirements of humankind and the psychological needs of subjects from whom this information is sought, we must offer a rational perspective from which decisions regarding the integration of these occasionally disparate needs are to be met.

SUBJECTIVE FACTORS: EMPATHY AND PROJECTION

The first step in approaching this judgmental balance is to attempt an exploration of the basis upon which "psychological well-being" is to be judged. Since we cannot simply ask our subjects their state of mind, or their emotional status, we will be forced to use some indirect measure or measures as the basis of our judgment. However, under these conditions there are several sources of potential distortion in formulating a response to this problem. There is, of course, the usual problem of anthropomorphism: that is, we as humans may attribute to animals abilities, functions, and emotions that we judge to be true of ourselves. Under the best of circumstances, when considering other members of our own society, we see this human quality most positively as a capacity to *empathize* with others—we can "put ourselves into their shoes." When validly applied, empathic response to another is clearly a highly valued characteristic. As we shall consider below, however, this apparent empathy may not always be validly applied.

When similar processes of interpersonal judgment are engaged in for less desirable, defensive needs of the observer, we see such activities as *projection*—in this case one's own motives or reactions are falsely attributed to another. In our human relationships or interactions, projection may occur, for example, as a response to our own psychological needs, perhaps because we hold such feelings ourselves but they are somehow unacceptable to us; or, as another example, we may project some attitude or motive to another, because we need to explain to ourselves why we react to a situation or to the other individual in the way that we do. Thus, we may feel frustration or anger, or feel helpless and manipulated by others in a particular situation, and though we may reject these negative or vulnerable feelings in ourselves, we unconsciously attribute these feelings to others, interpret their behaviors in this light, and react accordingly. Such projections may markedly alter the course of our interactions and basic relationships with others in our environment, often to the detriment of both parties. In any case, when projective assessments are made in this fashion, other, equally rational and moral individuals may fail to read into the behavior of another person those features that seem so clearly evident to the individual engaging in the projection. Seen another way, both of these processes of empathy and projection involve internal and subjective, rather than external and objectively verifiable, factors as sources of the judgments and the emotional reactions elicited in the observer of a given situation.

But in each of these human-to-human attributions, there is at least a reasonable basis for judging the comparability of someone else's capacities to one's own. The other person in a shared situation could be feeling the same emotions as those evoked in oneself—he or she could be "homesick" or content, enraged or pleased, bored or positively stimulated. But one need not be either a cognitive scientist or a psychoanalyst to recognize how often such attributions can be dead wrong (particularly, we may feel, when others incorrectly attribute certain qualities or reactions to us). As the history of cultural anthropology and of diplomacy have often demonstrated, and as so frequently happens when two human cultures intersect under any circumstances, members of one cultural group may misread or misinterpret the responses and hence the feelings and internal state of persons from a different cultural background. The wish to be empathic does not necessarily mean that we judge the feelings of the other person correctly. A laugh may imply joviality or embarrassment; tears may be those of joy as well as sadness. A street person may desperately feel the need for stable living quarters or may relish the sense of freedom to be found in this lifestyle, in ways unthinkable to the average person seeking to help improve matters.

"Why can't a woman be more like a man?" went Henry Higgins' refrain, capturing the problem of confused interpretation of the differing perceptions and feelings of men and women of even the same culture! We exist in a world in which nations and cultures often fail to understand the reactions of other groups, in which men and women within a culture are often helpless at predicting each other's reactions, and in which parents fail to understand what their children are thinking and vice versa. One cannot help but marvel at the fact that there are those among us who are unhesitatingly prepared to tell us what the members of other species perceive, feel, and understand about the world around them on the basis of their own intuitive sense. Surely we can accept the fact that humans may be at least as wrong, at least as often, in their interspecific attributions as they are in judging responses of those in the next country, the next city, the next room, or the next chair.

Indeed, even when we consider the most physically definable dimensions of the environment—light, colors, sounds, smells, space—neither our own perceptions nor our quantitative measurements of these physical features can be viewed as coincident with a given animal's perceived environment.

Obviously, this is not a new idea. Many decades ago, one of the pioneers of animal behavior research, von Uexkull, discussed the critical concept of the *Merkwelt* of each animal, the psychological, or perceived environment, *as experienced by the subject, as opposed to the environment as perceived from the perspective of the human*. This perceived "world" and the "effector" world towards which the subject acts, form a closed unit, the *Umwelt* as defined by von Uexkull. In discussing the concept of Umwelt, von Uexkull (1934) went on to say:

> The Umwelt of any animal that we wish to investigate is only a section carved out of the environment which we see spread around it—and this environment is nothing but our own human world. The first task of Umwelt research is to identify each animal's perceptual cues among all the stimuli in its environment and to build up the animal's world within them. . .
>
> We are easily deluded into assuming that the relationship between a foreign subject and the objects of his world exists on the same spatial and temporal plane as our own relations with the objects in our human world. This fallacy is fed by a belief in the existence of a single world into which all living creatures are pigeonholed. This gives rise to the widespread conviction that there is only one space and one time for all living things. (p. 14)

Thus, lacking verbal report and fully accurate intuitive attribution, have we no basis on which to make reasonable judgments regarding desirable qualities of the animal environment? Certainly we have! The question is, what forms of denotatively referenced elements of the animal's response or pattern of adjustment to its environment can be used to make our judgments, instead of those qualities that we might attribute to the animal and its presumptively perceived environment on the basis of our anthropocentric view.

OBJECTIVE FACTORS: "NATURALNESS" AND INDIVIDUAL CHARACTERISTICS

One approach that fails the test of rational assessment is to act on the notion that if an element is present in the natural state, it is desirable. It is clear that there is much in nature that few of us would see as necessarily enhancing well-being—predation, group rejection, toxic plants, or drought. Obviously not all environments are created equal, in the wild or anywhere else. Is it not the case that animals make choices of environments in the wild? Will not more dominant troops of monkeys choose to live in areas of food abundance and accessible water rather than in areas that require many more hours of foraging to meet dietary needs? Do they not choose to rest or sleep in less predation-risky settings? Is it not true that animals injure conspecifics in battles over rewards that might be trivial to humans, such as the chance to sit on a "better" rock? Have we not determined that animals choose to leave their natal groups and enter, at times with some difficulty, other social groups, while these same groups may make outcasts of other group members? Finally, is it not the fact that the processes of evolutionary change have worked continuously to diminish the misfitting of species to various elements of their natural environment, thus suggesting that the species is not "evolutionarily content" with all aspects of the current surround?

Unfortunately, in spite of the wide acceptance of these aspects of the natural history and biology of our subjects, the assessments of "desirable" environments by some well-intentioned observers seem ultimately to depend on a sense, a feeling if you will, of what *seems* right to us as humans. In addition to the fallacious perspective outlined by von Uexkull, these judgments may often be based on naive notions of a generalized environment within which a given species is presumed to live in nature and on the observer's personal attributions of the animal's response to those features. That is, as R. Nadler (personal communication, 1990) has pointed out, we are confronted with an aesthetic judgment, based on a complex of conscious and unconscious anthropomorphic attributions mixed with a "nature knows best" philosophy, rather than with judgments accessible to rational discourse and empirical validation.

One should consider in this regard whether concerned individuals from cultures having a different interplay with nature than our own would set the same priorities in determining the sources of well-being in captive primates. Consider, for example, a society in which food is always scarce, or one in which the community has not yet achieved the current western preoccupation with the intrinsic value of physical exercise and a slim waist. Would its members be as bent upon encouraging animals to work for their food and keep trim through induced daily exercise as are members of our own society? Unlike judgments based on materials derived from the scientific method, aesthetic judgments are inherently embedded in culture and in the individual observer. As has been said, "We never see the glass through which we look"; but if we are to achieve the goal of providing our primate subjects an environment in the laboratory that makes them feel better, rather than simply making some of us feel better, we must question the "givens" upon which these judgments are based. To fail to differentiate between aesthetic and rational decisions is potentially to doom us to a relentless and ultimately unresolvable conflict, in which both the animals and humankind will lose.

It follows from this viewpoint that the first basis of this rational approach is formally to recognize the very considerable differences that exist among different species of primates and often among classes and individuals within each species. Literally thousands of studies have demonstrated that species and families of primates clearly differ not only in size and appearance, but in their sensory, perceptual, cognitive, and affective characteristics as well. That is, imposition of guidelines or procedures that are valid in meeting specified goals for primates in general may be no better than a lowest-common-denominator approach or a straitjacket into which all species are forcibly squeezed. Thus, floor space may be of paramount concern to the terrestrial human, but height may be most comforting to an arboreal monkey. Close groupings may be positive for females of a species, but males might find similar spacing stressful. One species may prefer to live in relative isolation though most others are strongly gregarious.

To be sure, all primates have certain attributes which characterize members of the order Primates. As warm-blooded vertebrates, all primates hold in common a number of basic nutritional, respiratory, and thermal requirements. But even within these objectifiable, quantitatively measurable domains, important species differences exist. Thus, for example, certain lawful relationships hold between species of different sizes, which can be used to achieve particular goals in maintaining different species under nominally parallel conditions.

To take one example, according to our best data on the allometry of the nonhuman primates, the metabolic costs of moving through the environment differ as a function of body mass, as is true of other mammals. Metabolic rate does not change linearly with size, but rather with the .75 power of mass. Thus, a species twice the size of another does not incur twice the cost of moving a given distance within its environment, and if we establish as a goal the provision of each species with a comparably sized environment, this metabolic scaling factor must be taken into account. That is, a species twice the size of another needs proportionately more than twice the space as the first if their energy costs in utilizing that environment are to be made equal.

The same approach would similarly affect our decision regarding the effort we might wish to require of animals in obtaining their daily food, that is, the distances they must travel or the work they must perform in extracting a given caloric input from their environment. Since the costs of performing a given action differ from species of one size to species of another, the caloric input resulting from that action, should, if equality of treatment is a goal, differ in parallel fashion, or else the required work should be proportionately adjusted. Certainly there are other objectively assessable outcome measures that could be used as indirect reflections of comparable desirable environments for different species—general health (e.g., lack of disease, wounding, or quality of coat), longevity, reproductive rate, fetal loss, or rate of growth, to name several obvious choices.

But each of these quite overt measures must also be judged in terms of species-relevant parameters. Given the health problems and reproductive losses, which are repeatedly reported in field studies, we may or may not choose to set our laboratory criteria on the basis of our estimates of the characteristics of a given species in nature. But, however we choose to incorporate such elements in our determination of the quality of laboratory environments and the life of the primates living within them, at the least these types of measures afford an objective basis on which such judgments can be made.

Finally, at the individual level, all primates seek control over salient features of their environment (Umwelt), whether they are housed alone or in groups, in cages, arenas, or in the wild. Their ability to adjust to the changing demands placed on them, to alter aspects of the array of stimuli that they perceive as confronting them, and to organize their responses into strategies that are effective in producing the outcomes they seek, may be seen as engaging the animals' most complex cognitive and affective capacities. Depending on the criteria we employ, such efforts can be viewed as underwriting most effectively the development and sustenance of our subjects' psychological well-being. But given the interspecific variation in cognitive, perceptual, and motor capacities, "control over the environment" means something quite different for one species than for another.

It is fair to say that although all of those concerned with primate well-being may agree on the conceptual basis of the environmental control issue, the actual nature and hierarchy of desired dimensions of control for different species is unknown. As with each of the features of the environment regarding which we seek understanding in our efforts to provide a rational basis for the enrichment of nonhuman primate laboratory environments, only thoughtful, detailed, and scientifically valid empirical research can provide the answers we seek. But regardless of the particular elements for which we pursue understanding and objective judgment, we can be certain that each aspect of laboratory environments for primates must reflect an appreciation of common evolutionary trends within the order, as well as species-specific and individual perceptual or cognitive capacities and behavioral propensities. Only in this way can we find truly common ground on which to bring together those diverse members of our society who wish to find a means of advancing human welfare through primate research while maximizing the psychological well-being of the animals whose participation we require.

Reference

von Uexkull, J. (1934). A stroll through the worlds of animals and men; a picture book of invisible worlds. In Schiller, C.H. (1957). *Instinctive behavior* (pp. 5–80). New York: International Universities Press.

CHAPTER 7

THE ROLE OF INDIVIDUAL DIFFERENCES IN PROMOTING PSYCHOLOGICAL WELL-BEING IN RHESUS MONKEYS

STEPHEN J. SUOMI AND MELINDA A. NOVAK

There is little doubt that the Animal Welfare Act of 1985, which mandated that captive nonhuman primates be provided with "a physical environment adequate to promote psychological well-being. . .," has presented formidable challenges to researchers and regulators alike. The real doubt concerns how such promotion might be accomplished and, more fundamentally, how psychological well-being is to be operationally defined and ultimately evaluated for nonhuman primates.

Psychological well-being is a term that primate researchers and regulators have inherited from Congress, one that clearly goes far beyond the well-established boundaries of humane care and treatment of research primates dictated by previous versions of the Animal Welfare Act (1966, 1970, 1976). Although this new legislation has a worthy goal—the perception of animals as living organisms and not as mere research tools—implementation of the stated provision presumes specific knowledge and techniques that do not, in fact, currently exist. Instead, the force behind the amendment seems to stem more from anthropomorphic assumptions than from hard scientific evidence or extensive regulatory experience (Holden, 1988). Nevertheless, it appears that all involved will have to learn to live with these new requirements. Real questions remain, however, with respect to how these requirements might best be interpreted and then implemented.

Such questions are probably exacerbated by the fact that most primate researchers and regulators come from quite different traditions with respect to how compliance with legally mandated standards (or rules of scientific evidence) is best evaluated and optimally accomplished. The regulatory tradition, well reflected in the current United States Department of Agriculture (USDA) standards for the care and treatment of nonprimate animals in both research and commercial settings, has focused largely on *structural* aspects of care and treatment, that is, on features that can be measured in physical terms. Thus, standards have been developed that emphasize factors such as the physical dimensions of cages, the range of temperature, humidity, and number of air exchanges in housing rooms, or the frequency of feeding and cage cleaning. By and large, such standards have proven to be useful in generating guidelines that are oriented toward maintaining the physical health of the animals involved.

Most scientists who study primate behavior and physiology focus instead on *functional* aspects of their subjects or housing environments. As a result, these individuals are much more likely to be concerned with the actual "performance" of the animals, for example, the frequency and duration of certain behavioral actions, the levels of various hormones, or the patterns of specific physiological processes such as heart rate, electroencephalograph readings, or immune system responses. Of course, such measures are also germane to considerations of physical health and, in fact, may be used by attending veterinarians to monitor the health status of the animals under their care (see G.P. Sackett, chapter 5 in this volume, for a further discussion of these issues).

In summary, it appears that these two frames of reference—one emphasizing physical features of housing environments, the other emphasizing behavioral and physiological features of the animals living in such environments—have largely converged when physical health is the issue at hand, as has been the case in the past. By and large, this convergence has been successful. For example, primates maintained in captive environments typically outlive their feral counterparts by an enormous statistical margin (Novak & Suomi, 1988).

However, "psychological well-being" clearly involves far more than physical health, and it is here that two basic problems arise. First, there is a definite lack of consensus among scientists as to how psychological well-being might be defined in functional (operational) terms. Second, it is not entirely obvious how issues concerning psychological well-being should be translated into characteristics of physical environments that would have broad generality to all or most primate species, or to individuals within a given species. Each of these problems will be discussed in turn.

FUNCTIONAL CRITERIA FOR EVALUATING PSYCHOLOGICAL WELL-BEING

At this point, it is safe to say that there is no universally accepted way of defining psychological well-being for nonhuman primates among researchers, let alone among veterinarians, regulators, animal rights activists, or members of the general public. Instead, a wide range of specific definitions permeates the scientific literature, in many cases representing quite different points of view or even general frames of reference. For example, some definitions rely primarily on behavioral characteristics or the product of behavioral events such as successful reproduction or care of offspring (Snowdon, in press). Others focus more on physiological processes, on preestablished levels of certain hormones, or on changes in the immune system (see Coe, chapter 10 in this volume). Still others eschew any specific definitions of what constitutes psychological well-being and focus instead on what it is not (see Keeling, Alford, & Bloomsmith, chapter 8 in this volume).

Novak and Suomi (1988) have recently offered a different strategy for characterizing psychological well-being in nonhuman primates. Rather than providing a single definition, Novak and Suomi have suggested a number of dimensions along which specific criteria might be established for nonhuman primates. The first of these, *physical health*, is already relatively well established in the veterinary literature, at least for some species. The second dimension, the *behavioral repertoire* of the primate in question, also has a relatively firm basis in both the laboratory and field literature for many species, although what specific aspects of any individual's behavioral repertoire are relevant for defining psychological well-being is a question certainly open to potential argument (cf. Novak, O'Neill, Beckley, & Suomi, in press). The third dimension, the *degree of stress* encountered by the animal (or the relative *absence of distress*), can be characterized in either behavioral or physiological ways, or even in terms of specific features in the animal's living environment. Finally, Novak and Suomi have proposed that the manner and degree to which the animal copes with challenges encountered in its environment—that is, its *coping capabilities*—provide a meaningful dimension with which to evaluate its relative psychological well-being.

Of course, these four dimensions need not be mutually exclusive—indeed, Novak and Suomi (1988) have argued that under favorable circumstances, they should actually converge. Nor is the claim being made that these dimensions are exhaustive, that they encompass all the ways in which psychological well-being can be conceptualized or operationally defined. On the other hand, these four dimensions do provide a broad-based framework within which there may be alternative ways of characterizing and evaluating psychological well-being. Furthermore, each dimension can be defined in operational terms, which, in turn, can be used to generate standards or criteria against which any given nonhuman primate subject's appearance, behavior, level of physiological activity, or response to challenge can be compared or evaluated. Although a detailed discussion of the ways in which such translations to operational definitions can actually be carried out is beyond the scope of this chapter (cf. Novak et al., in press, for a more complete discussion of the relevant issues), the bottom line is that these dimensions can be used to generate "performance standards" for assessing psychological well-being in captive nonhuman primates.

If performance standards that characterize and evaluate psychological well-being can indeed somehow be developed, then the subsequent task of characterizing and evaluating environments as to their adequacy in promoting psychological well-being seemingly becomes simpler. Presumably, all one would have to do would be to evaluate whether the primates living in the particular environment under scrutiny were having their psychological well-being adequately promoted. Environments that housed such primates would be in compliance and those that didn't would not.

INDIVIDUAL DIFFERENCES IN REQUIREMENTS FOR "ADEQUATE" PSYCHOLOGICAL WELL-BEING

The scenario outlined above—using performance criteria for subjects to provide the basis for determining whether a particular housing environment adequately fosters their psychological well-being—seems straightforward and deceptively simple. However, there are real practical problems with this approach and, although not insurmountable, they complicate the scenario substantially. These problems arise from the fact that across the range of nonhuman primate species, there are enormous differences in health needs, behavioral repertoires, susceptibility to environmental stressors, and coping capabilities. Environmental standards based on the normative characteristics and requirements of one species may be grossly inappropriate for many other primate species. Such a conclusion is hardly surprising when one considers the vast range of differences in natural habitats to be found within the primate order, as well as the equally diverse social structures characterizing even taxonomically close species. Still, given appropriate knowledge of a species' range of natural habitats and species-normative social structures, it would seem possible to be able to "fine-tune" general standards in order to accommodate species-based variability in factors relevant for promoting psychological well-being. However, this consideration does not solve all the problems.

Within any given primate species, there are almost always major differences in health needs (e.g., standard diet), behavioral repertoires, degree of stress or distress, and coping capabilities for conspecifics who differ in age or sex. This fact suggests that the specific environmental requirements for promoting psychological well-being of infants, juveniles, adolescents, and adults of either sex may also differ in fundamental ways within at least some primate species. For example, prolonged separation from one's natal social group may have devastating consequences for the psychological well-being of a 5-month-old macaque (Sackett & Erwin, 1981), although prolonged separation from the natal social group may be essential for a pubertal male's very survival, let alone his psychological well-being (Simonds, 1965).

Still, age- and sex-based differences in environmental requirements for adequate promotion of psychological well-being need not pose insurmountable difficulties, at least based on data from some species. For example, rhesus monkeys living in feral troops differ dramatically as a function of age and sex in terms of which subset of possible social partners they actually spend most of their time with and devote most of their social actions to. Most of these smaller subgroups (or self-imposed, short-term isolation) can be readily modeled or simulated within laboratory settings (cf. Novak et al., in press). Thus, practical solutions to problems stemming from age–sex differences inherent in virtually every primate species are clearly possible, if not already largely in hand.

Nevertheless, there remains at least one additional problem of significant importance. An increasingly robust body of data has identified major differences between individuals of comparable species, age, sex, and (in some cases) rearing experience in terms of their response to stress. Recent studies of rhesus monkeys again provide data with provocative implications.

These studies have disclosed substantial individual differences in both behavioral and psychological response to novelty and challenge of the sort routinely encountered in feral environments. Some rhesus monkeys consistently respond to such relatively mild environmental stressors with unusual behavioral disruption and physiological arousal, including prolonged activation of the hypothalamic-pituitary-adrenal axis, increased monoamine turnover, heightened sympathetic nervous system activity, and abnormal immune-system response. The very same mild environmental stressors elicit trivial behavioral reactions and only transient physiological arousal in other members of the population. Such patterns of individual differences in response to stress have been found in both laboratory and feral populations of rhesus monkeys (Suomi, in press), and they appear to be highly heritable, although clearly subject to environmental influence or modification (Suomi, 1987). Moreover, in relatively stable physical and social settings, these differences are remarkably stable developmentally; rhesus monkey infants who display extreme behavioral and physiological responses to environmental novelty or challenge are also likely to continue to display extreme reactions as juveniles, adolescents, and even young adults in the absence of intervention (Suomi, in press).

The existence of such dramatic individual differences even among rhesus monkeys of the same age, sex, and rearing background poses real problems for designing living environments capable of fostering the psychological well-being of all their inhabitants. Although "high-reactive" rhesus monkeys constitute a minority of virtually all rhesus monkey groups screened to date for individual differences in response to challenge, that minority is nevertheless substantial, typically accounting for approximately 20% of every population examined thus far (Suomi et al., 1989). This large minority means that an environment that provides acceptable levels of novelty and challenge for the majority of its rhesus monkey inhabitants may, in fact, be highly stressful for the remaining individuals. The consequences for these latter individuals of living in such an environment might include difficulties in coping, displays of behavioral abnormalities, and even health problems, particularly if the stress involves long-term compromise of immune-system functioning. On the other hand, if the environment is designed to be so benign as not to trigger stressful responses in these high-reactive individuals, it may also turn out to be needlessly sterile, bland, or boring for the rest of the rhesus monkey population. It seems difficult, if not impossible, to create any single physical environment capable of promoting, in optimal fashion, the psychological well-being of all members of the population. Certainly, no such optimal environment exists in the wild—not even on Cayo Santiago.

IMPLICATIONS FOR DESIGNING "OPTIMAL" ENVIRONMENTS

The presence of dramatic individual differences among rhesus monkeys in their characteristic response to environmental stressors suggests, as argued above, that no single environment will be best for all conspecifics. Instead, it implies that some compromises will have to be made both by those who design and maintain captive environments for rhesus monkeys and by those given the regulatory responsibility of determining their adequacy for promoting psychological well-being. What is an optimal environment for rhesus monkeys? Is it one that maximizes the well-being for most of its inhabitants, even though certain individuals may be suffering in such settings? Or is it an environment that minimizes the stress experienced by the most highly reactive individual in the population, even though it may be sterile and boring for most of the other monkeys? Effective guidelines for making such decisions do not currently exist.

Moreover, the problems posed by the presence of individual differences in response to environmental novelty and challenge are unlikely to be limited to rhesus monkeys or even the genus *Macaca*. The literature is full of references to similar patterns of individual differences across a wide range of species including humans and nonprimate mammals (Kagan, Reznick & Snidman, 1988). Thus, the same problems are likely to emerge when designing environments capable of promoting psychological well-being for other primate species.

Indeed, similar problems have been encountered throughout history by those given the responsibility of promoting the well-being—psychological or other—

of human populations. The issue of creating optimal environments has been thoroughly debated by scientists, philosophers, and political leaders of every recent generation, and whatever consensus has emerged has usually turned out to be transient. The wisdom of Solomon was not sufficient to provide an optimal environment for all of his subjects. It may also be insufficient to create an optimal environment for all captive nonhuman primates.

References

Animal Welfare Act, 7 U.S.C. §§ 2131-2157 (1985).

Animal Welfare Act, Pub. L. No. 94-279, 90 Stat. 417 [Codified as amended at 7 U.S.C. §§2131-2156 (1976)].

Animal Welfare Act, Pub. L. No. 91-579, §2, 84 Stat. 1560 [Codified as amended beginning at 7 U.S.C. § 2131 (1970)].

Holden, C. (1988). Experts ponder simian well-being. *Science, 241*, 1753–1775.

Kagan, J., Resnick, J. S., & Snidman, N. (1988). Biological bases of childhood shyness. *Science, 240*, 167–171.

Laboratory Animal Welfare Act, Pub. L. No. 89-544, §1, 80 Stat. 350 [Codified as amended beginning at 7 U.S.C. § 2131 (1966)].

Novak, M. A., O'Neill, P., Beckley, S., & Suomi, S. J. (in press). Naturalistic environments for captive primates. In E. G. Gibbons, E. Menze, & E. Wyers (Eds.), *Indoor naturalistic habitats*. Stony Brook, NY: SUNY-Stony Brook Press.

Novak, M. A., & Suomi, S. J. (1988). Psychological well-being of primates in captivity. *American Psychologist, 43*, 765–773.

Sackett, G. P., & Erwin, J. (1981). Effects of weaning age on mortality of colony-born *Macaca nemestrina*. *Theriogenology, 15*, 505–511.

Simonds, P. E. (1965). The bonnet macaque in South India. In I. Devore (Ed.), *Primate behavior* (pp. 175–196). New York: Holt, Rinehart, & Winston.

Snowdon, C. T. (in press). The criteria for successful propagation of endangered species. *Zoo biology*.

Suomi, S. J. (1987). Genetic and maternal contributions to individual differences in rhesus monkey biobehavioral development. In N. Krasnegor, E. Blass, M. Hofer, & W. Smotherman (Eds.), *Perinatal development: A psychobiological perspective* (pp. 397–420). New York: Academic Press.

Suomi, S. J. (in press). Primate separation models of affective disorders. In J. Madden (Ed.), *Adaptation, learning, and affect*. New York: Raven Press.

Suomi, S. J., Scanlan, J. M., Rasmussen, K. L. R., Davidson, M., Boinski, S., Higley, J. D., & Marriott, B. (1989). Pituitary-adrenal response to capture in Cayo-derived M-troop rhesus monkeys. *Puerto Rico Health Sciences Journal, 8*, 171–176.

CHAPTER 8

DECISION ANALYSIS FOR DEVELOPING PROGRAMS OF PSYCHOLOGICAL WELL-BEING:
A BIAS-FOR-ACTION APPROACH

MICHALE E. KEELING, PATRICIA L. ALFORD, AND MOLLY A. BLOOMSMITH

Many conferences and workshops have been designed to provide a forum for developing definitions and measures of psychological well-being in nonhuman primates. Though discussing these issues in the abstract is appropriate, we want to propose complementing these discussions with the more aggressive approach of implementing clinical trial programs in psychological well-being.

Theoretical discussions aimed at developing definitions and measures of psychological well-being in nonhuman primates are valuable. At present, we feel that theorizing must give way to action. First, progress in this field has been excessively slow. Second, if we do not soon adopt a more aggressive approach and demonstrate significant advances in this area, our critics will continue to be successful in stampeding the public and our legislative bodies into developing additional cost-prohibitive regulations that may cripple our research programs.

Two concepts, one a means of applying the other, can help us to implement clinical trial programs in psychological well-being more aggressively. The first

This investigation was supported by research grant 1 R24 RR03578 from the Division of Research Resources, National Institutes of Health.

concept, a "bias for action," comes from the business community and was described in *In Search of Excellence* (Peters & Waterman, 1983). These authors were, of course, describing corporate America, but a bias for action can equally apply to the scientific community. In short, the absence of complete data should not necessarily subvert immediate action. Because overintellectualizing a problem can lead to paralysis, we must be predisposed to taking action, even though that action may be based on less information than we would like. Obviously such action cannot be careless and, consequently, must be based on a sound conceptual framework.

The second concept, critical decision analysis, can be helpful in providing such a framework. Decision analysis is a systematic approach that can use our accumulated knowledge and experience of animal well-being to make explicit, quantitative, and prescriptive value judgments.

With a bias for action and decision analysis, we can make better use of available data to implement immediate programs, specifically behavioral enrichment programs. Admittedly, these programs will be imperfect and subject to revision; however, given the current rate of progress and the outside forces impinging on our field, action is imperative. We will illustrate the value of this process by constructing a hypothetical decision-analysis tree using a menu of currently accepted behavioral enrichment techniques.

A BIAS FOR ACTION

In 1983, the nationwide best-seller *In Search of Excellence* (Peters & Waterman, 1983) presented an intense multiyear study of 43 of the most successful companies in America. Although these companies represented a wide range of activities from high technology to pure service, they had eight basic principles in common. One principle that applies to our situation is the bias for action. As this term implies, companies with a bias for action are predisposed to taking deliberate (though not careless) risk and to placing action above planning. Peters and Waterman concluded that all successful programs are characterized to a degree by these qualities. Although the conferences and numerous workshops that we have participated in over the past 3 years are necessary and valuable, we propose that these discussions be complemented with the adoption of a bias for action.

Peters and Waterman cautioned us concerning the lethargy and inertia that a complex problem can cause if the problem is overintellectualized or becomes too politicized. Successful programs can be stymied by overcomplexity; a balance of simplicity and complexity must be established. Currently, business leaders discuss the need for developing an experimenting mentality. Perhaps as scientists we need to revive our experimental nature and improve the innovative element of our environmental enrichment programs. We feel a degree of urgency in developing more innovative trial programs in environmental enrichment.

Table 1 charts the major events impacting captive chimpanzees in this country for the last 50 years. This table supports the urgency fueling our bias for action by illustrating the traditional glacial speed of these significant events. In 1965, several major chimpanzee holders started experimenting with outdoor social housing of chimpanzees. Although this type of management was initiated primarily to support behavioral investigations, its cost-effectiveness and benefits to quality of captive life and domestic breeding soon broadened its applications.

The University of Texas facilities were initiated in 1977 and completed in 1979, and initial program funding was provided by the Division of Research Services at the National Institutes of Health. This program is currently funded by the Division of Research Resources. Between 1977 and 1980, there was a flurry of activity and of publications proposing psychological enrichment concepts. We had yet to coin the buzzword "psychological well-being," but we all shared a concept of what we were pursuing.

In 1980, a National Institutes of Health Advisory Task Force was established to develop a national chimpanzee management plan. Five years of effort, during which we spent considerable time discussing what we now call psychological well-being, resulted in the endorsement of a national chimpanzee breeding program by the National Institutes of Health in 1985; the sequence of meetings addressing psychological well-being began in 1986.

The point is that 12 years ago the University of Texas initiated a program to satisfy most of the psychological needs of our current chimpanzee population. If we had waited for convincing empirical data or a consensus of definitions, we would be even further behind current societal expectations in this arena.

DECISION ANALYSIS

In proposing a bias for action, we have to address questions concerning the logic of initiating programs before we have defined psychological well-being in animals. How do you start a program with so many unknowns and uncertainties? There is an analytical technique that we think is uniquely suited for application to this problem. Decision analysis has been used in business for years, and we have seen it applied increasingly in the medical field over the last 15 years.

Clinical decision analysis is a systematic approach to decisionmaking under conditions of uncertainty. It is a cognitive process with which medical and veterinary clinicians are thoroughly familiar and which they exercise routinely (Fetrow, Madison, & Galligan, 1985; Howard, 1968; Raiffa, 1968; Salman, Dargatz, Kimberling, & Ellis, 1988).

Table 1

MAJOR EVENTS IN THE UNITED STATES IMPACTING CAPTIVE CHIMPANZEES

1930	Robert M. Yerkes begins investigations; establishes laboratories in Orange Park, FL.
1963	Goodall reports on wild chimpanzees.
1965	Outdoor corrals established at the Yerkes Regional Primate Research Center (RPRC), the Delta RPRC, and the Holloman Air Force Base Chimpanzee Colony.
1966	Arnhem Zoo established.
1966	Animal Welfare Act enacted.
1972-1975	Yerkes RPRC: Ossabaw Island, SC, Field Station, GA, Kingdoms Three Animal Park (formerly Lion Country Islands), GA; Stanford University Outdoor Primate Facility, CA: LEMSIP[a]: Lion Country Island, FL.
1973	U.S. signs CITES[b] treaty.
1977	The University of Texas completes facilities for resocialization and breeding at Science Park.
1978	Fritz & Fritz, Pfeiffer & Koebner publish on resocialization.
1979	Erwin, Maple, Mitchell, & Markowitz develop enrichment concepts.
1980	NIH[c] Advisory Task Force on National Chimpanzee Management Plan created.
1983	Hepatitis vaccine developed.
1984	Human Immunodeficiency Virus (HIV) Chimp Model developed.
1985	NIH endorses National Chimpanzee Breeding Program.
1985	Psychological well-being amendments to the Animal Welfare Act enacted.
1986	National Chimpanzee Breeding Program Cooperative Agreements funded by Division of Research Resources.
1986	ASP[d] Meeting and APV[e] Workshop on psychological well-being conducted.
1986	AALAS[f] Symposium on Psychological Well-Being of Primates conducted.
1986	Goodall publishes *The Chimpanzees of Gombe*.
1988	ACLAM[g] Forum Conference on Psychological Well-Being conducted.
1988	Twentieth Annual Lab Animal Medicine Conference on Psychological Well-Being of Primates, University of Cincinnati, conducted.
1988	Conference on the Psychological Well-Being of Captive Primates, Harvard Medical School, conducted.

[a]LEMSIP = Laboratory of Experimental Medicine and Surgery in Primates
[b]CITES = Convention on International Trade of Endangered Species
[c]NIH = National Institutes of Health
[d]ASP = American Society of Primatologists
[e]APV = Association of Primate Veterinarians
[f]AALAS = American Association of Laboratory Animal Science
[g]ACLAM = American College of Laboratory Animal Medicine

The book *Clinical Decision Analysis* (Weinstein & Fineberg, 1980) provides an excellent description of the intricacies of this framework for making better decisions. Clinical decision analysis is a systematic way of deciding how to treat a serious clinical illness when you do not have the luxury of being able to wait for more information. In its simplest form, it is a common-sense approach to developing a working diagnosis by (a) considering alternatives and unknowns; (b) utilizing accumulated knowledge, experience, and judgment; and (c) balancing risks. It can be formalized to the extent of using additional analytical tools such as statistics, epidemiology, economics, clinical knowledge, and humanistic or humane concerns. We will illustrate this process by reviewing a hypothetical decision-analysis tree that concerns deciding whether to introduce a psychological enrichment program or wait until regulations force such an introduction.

THE DECISION-TREE MODEL

Figure 1 shows a proposed menu of currently accepted enrichment techniques. Enrichment in all of the five major categories may not have application in your individual program; however, with this much choice, some options should be available to everyone.

The decision-tree model illustrated in Figure 2 is the suggested way of visualizing the logical sequence to follow in problem solving. Weinstein and Fineberg use a small square to indicate a point at which the decisionmaker can elect one of several alternative courses of action—the decision node. A small circle indicates a point at which one of several possible events beyond the control of the decisionmaker may take place—the chance node. This decision tree is built on the strengths and weaknesses of electing to introduce a program designed to promote psychological well-being.

In keeping with our bias for action, we will extend the tree only in the direction of introducing a program, realizing that a similar tree could be developed for the option of waiting for the pending regulations. At the first decision node (Introduce Program), we can employ any combination of the five major enrichment categories (Figure 1), producing favorable or unfavorable results. Based upon the individual characteristics of the institution and its research programs, a tree can be developed to consider the favorable and unfavorable possibilities in each category where enrichment can be considered.

Figure 3 shows some of the possible favorable and unfavorable results that should be considered with regard to socialization. The other major categories can be similarly analyzed.

If there is insufficient information to determine a favorable or unfavorable result at any particular point, that branch of the analysis tree can be expanded as shown in Figure 4.

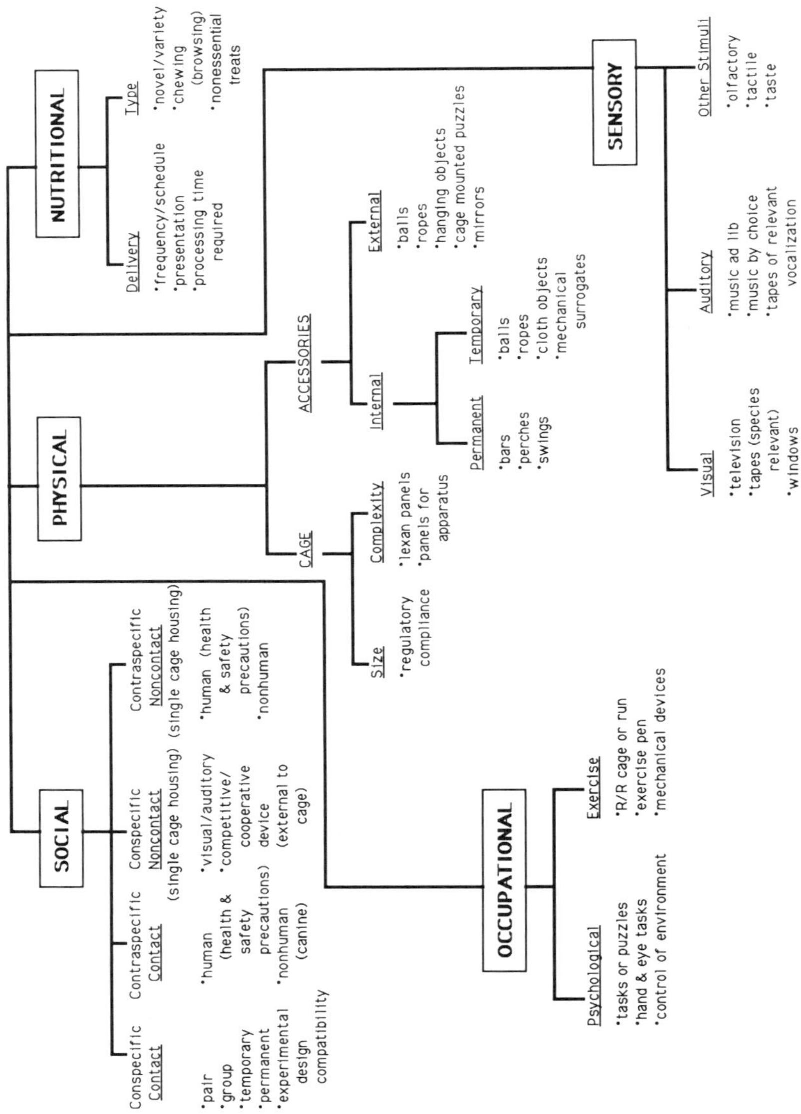

Figure 1 Captive primate enrichment options.

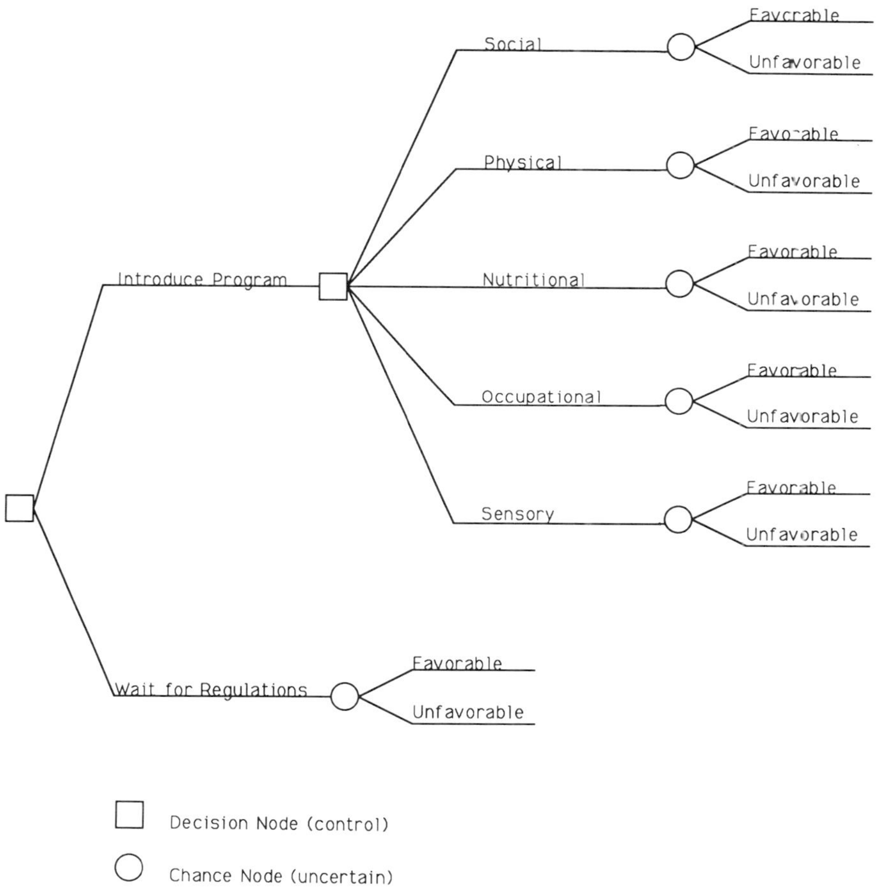

Figure 2 A decision tree incorporating five categories of environment-enrichment options as a base for introducing psychological well-being programs.

This "fanning" of the branches provides an opportunity for an additional decision node (small square) and a plan of action for seeking additional information before making decisions. This process can rapidly cascade to the limits of resources, but should be pursued as far as is practicable. We should all be making some effort to collect and disseminate information and experiences in this embryonic discipline.

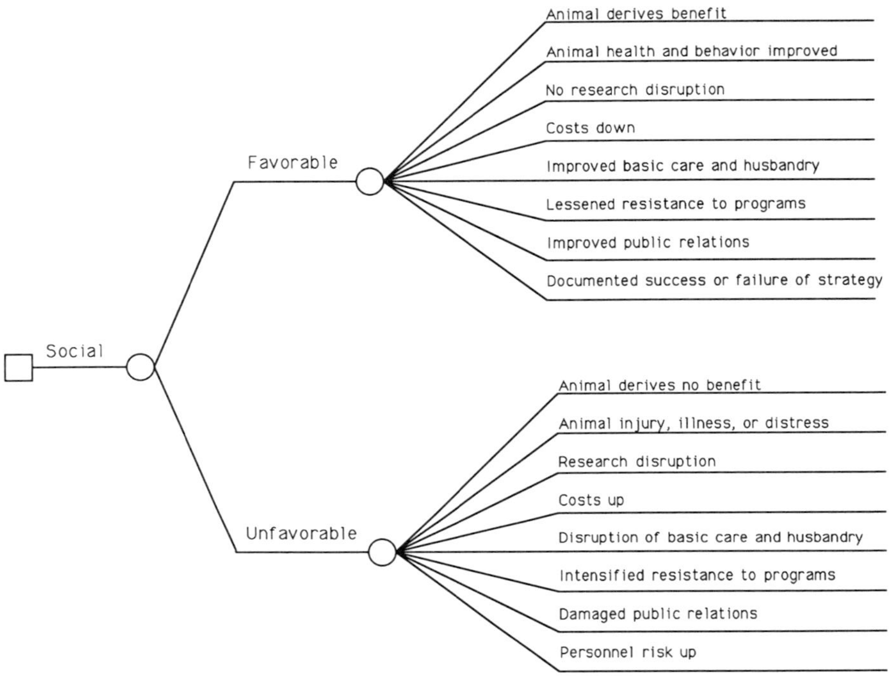

Figure 3 A decision tree outlining favorable and unfavorable results of providing social housing for nonhuman primates.

CONCLUSIONS

In conclusion, we would like to provide the following take-home message:

1. Go back to your institution with a bias for taking action concerning the introduction of programs designed to benefit the psychological well-being of primates.
2. Develop a written, custom-designed program for your facility. Get investigators and animal-care-and-use committees involved. The program can be designed around the individual characteristics of your facility's research program capabilities and limitations. Your program will probably complement the pending regulations, and your institution and the animals in it may benefit by having a head start.
3. Be very deliberate in developing this program, carefully weighing the risks and benefits. Employ decision-analysis techniques and make them as explicit, quantitative, and prescriptive as possible. In the medical field, decision analysis is a method of formalizing the combined objectives of doctors, patients, and society (Weinstein et al.,

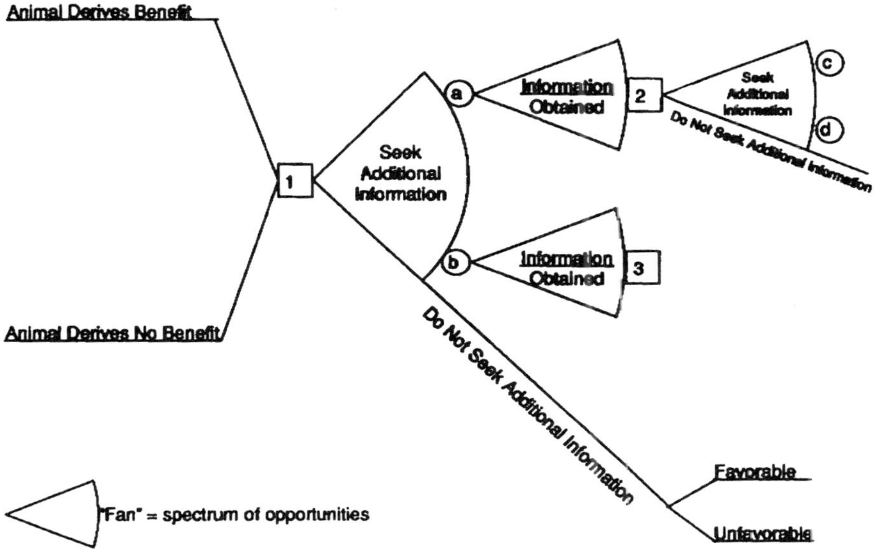

Figure 4 A decision tree illustrating "fanning" as a method of obtaining additional information.

1980). Biomedical research can employ this technique to achieve the combined objectives or needs of scientists, animals, and society.

References

Fetrow, J., Madison, J. B., & Galligan, M. A. (1985). Economic decisions in veterinary practice: A method for field use. *Journal of the American Veterinary Medical Association*, *186*, 792–797.

Howard, R. (1968). The foundations of decision analysis. *IEEE Transactions on Systems, Science and Cybernetics*, *SSC-4*, 211–219.

Peters, T. J., & Waterman Jr., R. H. (1983). *In Search of Excellence*. New York: Warner Books, Inc.

Raiffa, H. (1968). *Decision analysis: Introductory lectures on choices under uncertainty*. Reading, MA: Addison-Wesley Publishing Co., Inc.

Salman, M. D., Dargatz, D. A., Kimberling, D. V., & Ellis, R. P. (1988). An economic evaluation of various treatments for contagious foot rot in sheep, using decision analysis. *Journal of the American Veterinary Medical Association*, *192*, 1995–2004.

Weinstein, M. D., & Fineberg, H. V. (1980). *Clinical decision analysis*. Philadelphia: W.B. Saunders Co.

PART THREE

POTENTIAL PARAMETERS OF PSYCHOLOGICAL WELL-BEING

CHAPTER 9

THE SOCIAL NATURE OF PRIMATES

FRANS B. M. DE WAAL

EXTREME STANDPOINTS AND THE ROLE OF MODERATES

Within a logically consistent moral framework, there can be no gray areas or borderline cases. Admittance of a single exception would cast doubt on the rationale for all other choices. This "domino effect" is well understood by philosophers such as Singer (1976), Regan (1984), and others, who have proposed the moral equality of all species capable of suffering—from fish to human—and have condemned any use of animals, whether for food, sports, clothing, research, or other purposes. It is not evident, however, to what extent human systems of morality are or need to be logically consistent. One would think that there would be no reason for the emotionality of the current debates about abortion, euthanasia, and the treatment of animals if morality were just a matter of reasoning.

Whereas the animal rights movement can be credited with increasing awareness of the ethical issues involved in animal experimentation, its radical stance prevents a constructive contribution now that we have reached the stage of practical solutions. Clearly, the most vocal activists are abolitionists rather than regulationists of animal research. This extremism leaves the task of solving the ethical dilemmas to moderates within both the scientific community and the animal rights movement. By moderates, I mean people who recognize a need for animal research—hence our right (many would say obligation) to conduct such research—

I am grateful to Mary Schatz and Jackie Kinney for typing the manuscript, and to Lesleigh Luttrell for comments. Writing was supported by National Institutes of Health grant RR-00167 to the Wisconsin Regional Primate Research Center. This is publication number 28–032 of the WRPRC.

and also recognize a moral responsibility toward animals. Because moderates seek a balance between compassion for animals and the interests of humanity, it is inevitable that their moral philosophy should include gray areas. Although there can be no doubt that animals are capable of suffering, and that their suffering should be minimized, the prospect of alleviating human suffering through medical progress is given at least an equally prominent place in the moral equation.

Most moderates will agree that the time has come for drastic changes in the living conditions of laboratory animals. It is unfortunate, therefore, that the uncompromising ideology of the animal rights movement is matched by a rigid adherence to the old ways by many researchers. Yet, several research institutions are making progress, serving as pioneers for the rest, and a more open exchange of opinions and solutions seems to be under way. My contribution to this debate is the advocacy of social housing through a review of the intricacies of primate social life.

I find it hard to look at a singly caged monkey without getting depressed. If this empathic judgment sounds like groundless subjectivism, it is good to realize that for many years funding agencies and scientific journals implicitly agreed with it. Following Harlow's pioneering work on rhesus monkeys, social isolation and temporary separation from companions have been common research paradigms for a generation of experimental psychologists (Harlow, 1974; Hinde & Davies, 1972; McKinney, Suomi, & Harlow, 1971; Suomi, 1973). This is not the place to review the vast literature in this area; suffice it to note that it abounds with terms such as "depression," "terror," "despair," "frustration," and "distress."

When Kaspar Hauser first saw the daylight after many years of total isolation, he appeared to hear without understanding, to see without noticing, and to move his feet without walking. The boy's fate was considered a "crime against the human soul" (von Feuerbach, 1832). Because this judgment is how we view the social deprivation of a member of our own species, and because scientists have found similar behavioral pathologies in nonhuman primates kept under the same conditions, it is certainly not going too far to conclude that isolation of laboratory primates should never be imposed on a routine basis. Social deprivation should not be considered any more normal than, say, water or food deprivation.

PRIMATE SOCIAL SOPHISTICATION

Rhesus monkeys, baboons, chimpanzees, and other common laboratory primates are social beings. Their psychology has been shaped by millions of years of social evolution. These primates show the entire range of behavior necessary for survival in a group, from intense competition to close affiliation. They are capable of fine distinctions between kin, friends, and rivals, treating each category in a different manner. They can be ambivalent, being simultaneously friendly and aggressive to the same individual. They may try to move up the social scale and are opportunistic enough to form temporary alliances for this purpose. In short, these primates are as well-adapted to group life as fish are adapted to water.

Three examples of the social sophistication of nonhuman primates serve to highlight this point: political strategies, reconciliation behavior, and kinship bonds. These phenomena can be observed both in large captive groups and in natural populations. This is not to suggest that laboratories should keep their primates under naturalistic conditions, but investigators should at least be aware what they are depriving their subjects of if they decide that life in large groups is incompatible with their research objectives. They may wish to consider the compromise of housing their animals in small groups of three or four individuals.

Political Strategies

The dominance hierarchy among male chimpanzees is expressed in ritualized encounters in which the subordinate bows and grunts for the dominant. Subordinates may literally grovel in the dust. So long as ranks are clearly acknowledged, relations are relaxed. Serious problems arise when a male stops behaving submissively to one of the dominants. This in itself, without any overt provocation, is perceived as a challenge. The dominant begins to act more and more assertively, trying to force the other back into submission. If he succeeds, his rank is reestablished and the relationship relaxes again.

Instability persists, however, if the previous subordinate reacts with protest to the increased pressure. He may even begin to perform intimidation displays on his own initiative in order to test the dominant's nerves. With all his hair on end, he will hoot loudly at the dominant, throwing sticks and stones from a distance. This situation may develop into a full-blown power struggle, upsetting the entire chimpanzee community for months. Power struggles take a long time, because the entire network of relationships is tested in the process, and opportunistic shifts in allegiance further complicate matters. Obviously, physical strength and agility play a role, but the most important factor is the support both rivals receive from third parties. Just as in human politics, power struggles are largely decided by coalitions (de Waal, 1982).

Chimpanzees have a whole repertoire of vocalizations and gestures to recruit supporters. During a fight between males A and B, A may approach female C to put an arm around her shoulders, or he may scream at her, begging with outstretched hand. In the meantime, male B may hurry to his buddy, D, and briefly mount him, after which they jointly charge at A. One can imagine how complex things become if all members of a large group are drawn into the fray. It requires a high level of social intelligence and experience to exploit the mechanism of coalition formation to its maximum. The male who is best at it will be the winner of the contest.

That chimpanzees understand what they are up to during these group processes is suggested by so-called separating interventions. A male whose dominance rests on a coalition with another male will at almost any cost prevent contact between his partner and his rivals. If his partner sits with a rival, the dominant will begin a noisy display in front of them. Obviously, he cannot break up the contact by force, because such an attempt might lead to the other two banding together in defense. The dominant will display until the two separate, after which he will groom his partner to affirm their bond. By giving in to the pressure, the partner demonstrates that he values a good relation with the leader more than one with the rival. Hence, each successful separation confirms the existing coalition. I have witnessed hundreds of separating interventions; they are highly predictable if one knows the coalition structure of a group of chimpanzees. Such divide-and-rule tactics reflect awareness in the apes that a coalition excluding oneself is not in one's interest, and that formation of such coalitions needs to be nipped in the bud.

Reconciliation Behavior

Over the past decade, my research team has documented reconciliation behavior in two macaque and two great ape species (reviewed by de Waal, 1989a; & de Waal, in press). Other investigators have confirmed this phenomenon for two more primate species (York & Rowell, 1988; Cords, 1988). The general design of reconciliation studies has been a comparison between the behavior immediately following aggressive incidents and the behavior during control observations. These comparisons have demonstrated that former adversaries are "attracted" to one another. The probability of friendly contact increases following aggression, and this increase is selective. That is, individuals do not seek contact in an indiscriminate manner, but remember with whom they have fought and preferentially make contact with this party. This finding stands in contrast to the traditional view of aggression as a distance-increasing, dispersive mechanism.

Reconciliation takes a different form in each species. Chimpanzees kiss and embrace; bonobos (or pygmy chimpanzees) show a variety of sexual patterns, and macaques groom, or engage in hip-clasping. As indicated by the term *reconciliation*, the primary function of these contacts is assumed to be the restoration of social relationships. A finding consistent with this hypothesis is that postconflict reunions are most common in relationships that appear to have high social value. Thus, relatives (e.g., mothers and offspring) and coalition partners reconcile a greater proportion of their aggressive conflicts than individuals with loose social ties.

Even third-party mediation is known. Following an aggressive incident, two male chimpanzees may sit down at a distance from each other without proceeding to a reconciliation. Such a deadlock situation of mutual orientation without eye-contact may last up to half an hour. A female may break the ice by approaching one of the males and briefly grooming him, after which she will walk over to the other male, closely followed by the first male. In case this male fails to follow, she will return to pull at his arm. Having brought the rivals together, the female will withdraw (de Waal & van Roosmalen, 1979).

Kinship Bonds

Cercopithecine monkeys, such as macaques and baboons, are characterized by an elaborate matrilineal kinship network. In the wild, males normally leave the social group at puberty, whereas females stay together for life and form transgenerational bonds with maternal relatives. For example, one old female in our study group at the Wisconsin Primate Center would spend much of her time with her great-granddaughter, holding and carrying the infant whenever its mother left it behind. Occasionally, the mother would drop the infant off near the "babysitter" (i.e., loosen the infant from her ventrum while standing in front of her grandmother) before going her way.

Matrilineal systems were discovered in the 1950s by Japanese primatologists and have since been amply documented in a great variety of species (reviewed by Gouzoules & Gouzoules, 1987). Daughters generally "inherit" dominance positions right under their mothers in the hierarchy. The factors underlying rank inheritance are still unknown, but it is generally believed that genetic explanations do not suffice. It has been suggested that rank inheritance depends on coalitions among kin. Females show a strong tendency to support relatives in fights, which allows juveniles to defeat females older and bigger than themselves with the support of adult relatives (de Waal, 1977; Kaplan, 1977; Massey, 1977). Social explanations of rank inheritance are supported by experimental manipulation of the relative size of matrilines; drops in rank can be produced by the removal of key members of an individual's matriline (Chapais, 1988).

Not surprisingly, in view of the pivotal importance of female kinship relations, remarkable social cognition is applied in this domain. Experiments have shown that monkeys can successfully match color slides of familiar conspecifics on the basis of matrilineal relatedness between the depicted individuals (Dasser, 1988). A practical application of this knowledge has been demonstrated for wild vervet monkeys by Cheney and Seyfarth (1986). Long after aggression occurred between members of different matrilines, an increased probability of fighting remained between other members of the same two matrilines. Tensions between the initial combatants appeared to have spread to their respective kin. These results suggest that monkeys not only distinguish their own kin from nonkin, but also recognize kinship bonds among other members of their group.

IMPLICATIONS FOR THE LABORATORY

The above examples demonstrate that social companionship provides nonhuman primates with more than body warmth and affection. A social environment presents opportunities to develop complex social skills. It requires the animals to monitor variations in moods and social context, to respond to conspecifics in an appropriate manner, and to apply their impressive cognitive capacities. Social life challenges the animals, and induces moments of excitement and stress as well as subsequent relaxation. Nothing matches group life when it comes to breaking the cycle of boredom in a manner consistent with natural needs and inclinations.

The term *stress* is used above on purpose. Negative experiences are part of any truly variable environment. The interplay between negative and positive experiences, rather than positive experiences alone, is probably crucial for psychological well-being (see G. P. Sackett, chapter 5 in this volume). Within social groups, the primary negative experience is the receipt of aggressive behavior. Behavioral scientists need to educate the public about this aspect; monkeys and apes are not always the cute, playful creatures many people believe them to be.

Managers of primate colonies should do everything to prevent violence, but the only guaranteed way to eliminate it is individual housing. Trauma due to aggression is a price to be paid for social enrichment of the life of laboratory animals. In the natural habitat, aggression is quite common and occasionally results in fatal injuries. More important than these rare incidents, however, is the continuous competition over access to food; dominant animals often prevent subordinates from feeding. This produces differential survival rates for high- and low-ranking group members. For example, close to 90% of the individuals born into a wild toque macaque population failed to reach adulthood due to "socially imposed mortality" (Dittus, 1979). Needless to say, this form of stress is virtually absent in captive groups with their abundance of food and much lower mortality rates.

Fatalities due to fighting in primate colonies have led to charges of inadequate housing conditions and excessive crowding, but the relationship between spatial conditions and aggression is far from clear-cut. In a recent review, de Waal (1989b) compared studies of three different populations of food-provisioned rhesus monkeys: a free-ranging population near a temple in Nepal (Teas, Feldman, Richie, Taylor, & Southwick, 1982); a large group in an outdoor corral at the Field Station of the Yerkes Primate Center (Bernstein, Williams, & Ramsay, 1983), and another large group, with a higher population density, in a pen at the Wisconsin Primate Center. Contrary to predictions from crowding models developed on the basis of rodent studies, the aggression rates of female monkeys turned out to be almost identical under the three conditions, and the rate of male aggression was highest in the free-ranging population. Apparently, confinement has little effect on the aggression level in well-established monkey groups that have had time to adjust socially to their environment.

Social housing of nonhuman primates has three potential advantages that may appeal even to scientists and administrators unconvinced that living conditions need to be improved. First, replacement of the endless rows of stainless steel cages and access alleys by a number of large pens would result in a more effective use of the available laboratory space and reduce maintenance costs. Second, because reproductive abilities, such as appropriate copulation postures and adequate maternal skills, are learned automatically in a social environment, artificial insemination and hand-rearing would become superfluous except for special purposes. Finally, psychological and physical well-being are possibly interlinked (Coe, chapter 10 in this volume). If so, physiological, immunological, and neurological measures collected on isolation-reared (and hence psychologically deviant) nonhuman primates might not be representative, and therefore might be suboptimal for the development of models applied to the human species. Social housing would avoid that possibility.

Compared to the difficulty of creating a challenging and variable physical environment in the laboratory, creating a stimulating social environment is a fairly simple proposition. Group life is such an obvious, effective, and inexpensive way to eliminate apathy in laboratory primates that we should make it a goal to provide all of them with such a life in the future. To do so would imply that

1. Newborn primates should be raised by their mothers, because hand-reared individuals do not adjust well to social life. If hand-rearing is required, contact with conspecific age-mates should be provided.
2. Social isolation should be limited to short periods of experimentation.
3. If longer periods of isolation are necessary, contact with familiar conspecifics should be made possible, for example, by positioning companions within the audiovisual range of a chaired individual (see V. Reinhardt, chapter 11 in this volume), or by electronically established audio or video contact. In this way, social bonds can be maintained, and some degree of social comfort can be provided to the experimental subject.

Finally, we should not forget the fortunate circumstances that primates are intelligent and trainable. Innovative scientists will be able to develop procedures compatible with social housing for much of the research presently considered to justify individual housing. Processing (without sedation) of a large number of group-living monkeys for blood drawings, for example, is no problem for experienced animal caretakers; such processing does not need to require more than

1–2 minutes per subject. Practical details of successful procedures, many of which are already being applied on a small scale, should be communicated across institutions so that procedures that respect the social nature of nonhuman primates become the norm rather than the exception.

References

Bernstein, I. S., Williams, L., & Ramsay, M. (1983). The expression of aggression in Old World monkeys. *International Journal of Primatology*, *4*, 113–125.

Chapais, B. (1988). Rank maintenance in female Japanese macaques: Experimental evidence for social dependency. *Behaviour*, *104*, 41–59.

Cheney, D., & Seyfarth, F. (1986). The recognition of social alliances by vervet monkeys. *Animal Behaviour*, *34*, 1722–1731.

Cords, M. (1988). Resolution of aggressive conflicts by immature long-tailed macaques (*Macaca fascicularis*). *Animal Behaviour*, *36*, 1124–1135.

Dasser, V. (1988). A social concept in Java monkeys. *Animal Behaviour*, *36*, 225–230.

Dittus, W. (1979). The evolution of behaviors regulating density and age-specific sex ratios in a primate population. *Behaviour*, *69*, 265–302.

von Feuerbach, A. R. (1832). Kaspar Hauser: Beispiel eines Verbrechens am Seelenleben des Menschen. Zentralantiquariat D.D.R., Leipzig, 1983.

Gouzoules, S., & Gouzoules, H. (1987). Kinship. In B. Smuts, Cheney, D.L., Seyfarth, R. M., Wrangham, R. W., & Struhsaker, T. T. (Eds.), *Primate societies* (pp. 299–305). Chicago, IL: University of Chicago Press.

Harlow, H. F. (1974). Induction and alleviation of depressive states in monkeys. In N. White (Ed.), *Ethology and psychiatry* (pp. 197–208). Toronto, Canada: University of Toronto Press.

Hinde, R. A., & Davies, L. (1972). Removing infant rhesus from mother for 13 days compared with removing mother from infant. *Journal of Child Psychology and Psychiatry*, *13*, 227–237.

Kaplan, J. (1977). Patterns of fight interference in free-ranging rhesus monkeys. *American Journal of Physical Anthropology*, *47*, 279–288.

Massey, A. (1977). Agonistic aids and kinship in a group of pigtail macaques. *Behavioral Ecology and Sociobiology*, *2*, 31–40.

McKinney, W. T., Suomi, S. J., & Harlow, H. F. (1971). Depression in primates. *American Journal of Psychiatry*, *127*, 1313–1320.

Regan, T. (1984). *The case for animal rights*. Berkeley, CA: University of California Press.

Singer, P. (1976). *Animal liberation*. Jonathan Cape, London.

Suomi, S.J. (1973). Repetitive peer separation of young monkeys: Effects of vertical chamber confinement during separations. *Journal of Abnormal Psychology*, *81*, 1–10.

Teas, J., Feldman, H., Richie, T., Taylor, H., & Southwick, C. (1982). Aggressive behavior in free-ranging rhesus monkeys of Kathmandu, Nepal. *Aggressive Behavior*, *8*, 63–77.

de Waal, F. B. M. (1977). The organization of agonistic relations within two captive groups of Java-monkeys (*Macaca fascicularis*). *Zeitschrift für Tierpsychologie*, *44*, 225–282.

de Waal, F. B. M. (1982). *Chimpanzee politics*. London: Jonathan Cape.

de Waal, F. B. M. (1989a). *Peacemaking among primates*. Cambridge, MA: Harvard University Press.

de Waal, F. B. M. (1989b). The myth of a simple relation between space and aggression in captive primates. *Zoo Biology Supplement*, *1*, 141–148.

de Waal, F. B. M. (in press). Reconciliation among primates: A review of empirical evidence and theoretical issues. In W. Mason & S. Mendoza (Eds.), *Primate social conflict*. New York: Liss.

de Waal, F. B. M., & van Roosmalen, A. (1979). Reconciliation and consolation among chimpanzees. *Behavioral Ecology and Sociobiology*, *5*, 55–66.

York, A., & Rowell, T. (1988). Reconciliation following aggression in patas monkeys, *Erythrocebus patas*. *Animal Behavior*, *36*, 502–509.

CHAPTER 10

IS SOCIAL HOUSING OF PRIMATES ALWAYS THE OPTIMAL CHOICE?

CHRISTOPHER L. COE

In 1985, Congress passed the Improved Standards for Laboratory Animals Act (U.S. Congress, House, 1985), which amended and extended many provisions of the earlier Animal Welfare Act passed in 1966. Although a number of significant changes were proposed in the amendments, certainly one of the most controversial was the mandate to develop regulations to ensure "a physical environment adequate to promote the psychological well-being of primates" (Sec. 1752). In response to this amendment, the U.S. Department of Agriculture (USDA) Animal and Plant Health Inspection Service (APHIS) spent over 3 years soliciting and evaluating numerous comments on the proposed changes before presenting new regulations for the housing of nonhuman primates. Despite the scale and sincerity of the effort, there is still considerable concern that the proposed regulations may have a number of flaws. In addition, there is serious apprehension about the feasibility and cost of many of the new requirements for housing nonhuman primates. In this paper, we will consider potential problems in one particularly important aspect of these requirements—the emphasis on the social housing of all research subjects.

For the most part, under natural conditions, nonhuman primates are gregarious species that spend almost all of their lives in social groups. It is reasonable,

The research described in this paper was supported in part by National Institute of Mental Health Grant MH41659 and Office of Naval Research Contract N00014-87-0227. Several of the early hormone studies were conducted in collaboration with S. Levine, S. Mendoza, E. Smith, and J. Davidson at Stanford University, and were supported by NIMH Grant MH23645 to SL.

therefore, to believe that they should be provided with social companions whenever possible. Certainly, the housing of two monkeys together is one of the most straightforward and reliable means of enriching the day-to-day life of a research subject maintained in captivity. Why, then, would anyone be concerned about the possible impact of this aspect of the new regulation on primate research? The reason is that a simplistic and inflexible application of the requirement may lead to serious management problems at some primate facilities and may also have an adverse effect on certain types of research. A few of the complex issues involved in the mandate for social housing will be considered below: the management and husbandry implications, the possible effects on some types of research, and the financial impact.

IMPLICATIONS FOR COLONY MANAGEMENT AND HUSBANDRY

At first glance, it might seem that the new regulations for the social housing of nonhuman primates could have only a positive effect on the management and husbandry practices in most primate colonies. After all, boredom would be reduced and along with it the locomotor stereotypes that sometimes develop in individually housed primates. But the true complexity of managing a primate facility and caring for monkeys must be considered. It is no small task to choose the optimal social housing conditions for several hundred monkeys, many of whom may be maintained in a particular facility for their entire life span of 20–30 years. For example, not all pairs or social groups are compatible. Although some monkeys may benefit and thrive in the presence of social companions, there will be an increased risk of aggression and wounding among others. Especially when new pairs are formed and dominance relationships are being established, there is a strong likelihood that the veterinarians will be kept quite busy suturing wounds.

In this regard, considerable effort and close observation of the monkeys will be required to ensure that the psychological well-being of the *subordinate* monkeys has truly been enhanced. In most appropriately formed pairs and social groups, this will not be a major issue. However, in a small but significant number of the pairings, the subordinate monkey may be adversely treated by his dominant companion. Anyone who has seen how some dominant monkeys can relentlessly harass a subordinate monkey knows how serious a problem this can be. And it should be emphasized that this type of harassment and aggression is not just an artifact of captivity. It is well documented in studies of primates under natural conditions, ranging from the baboon in Amboseli to the chimpanzee in Gombe National Park (Smuts, Cheney, Seyfarth, Wrangham, & Struhsaker, 1986). In addition to monitoring the monkeys closely at the time of group formation, careful attention will have to be paid to the types of monkeys housed within the immediate vicinity of the monkey's cage. As we will discuss below, the inadvertent introduction of a female monkey into a room of previously acclimated male pairs may dramatically transform their behavior from affiliative to outright aggressive. This transformation can occur within hours after introduction of a female monkey to a room of males that have been living together peacefully for many months.

It should also be noted that one can provide at least as much evidence for a possible negative effect of social housing on physiology as for the proposed gain over individual housing. For example, if a socially housed subordinate female monkey shows an inhibition of menstrual cycles or impaired reproductive success, should that be considered to be a significant negative consequence of group living? Similarly, if a male squirrel monkey fails to show a seasonal increase in testosterone levels because he is the lowest ranking male in his group, is that a cause for concern? Some may argue that this line of reasoning is really a diversion from the major issue, which is, shouldn't we find better ways to house monkeys than to maintain them in individual cages? However, I would argue that the examples given above exemplify the real problems that managers of primate facilities will have to face. Further, the examples indicate the lack of consensus about how to measure our success at optimal management (Novak & Suomi, 1988).

For example, should one component of our measurement be reproductive success (e.g., the absence of complications of pregnancy or the number of viable offspring raised to maturity)? (See L. Johnson, A. Petto, & P. K. Sehgal, chapter 11 in this volume.) In the case of a developing primate, should weight gain, rate of behavioral maturation, and timing of puberty be considered, in addition to our subjective measures of psychological well-being? As most primatologists know, social housing in certain types of monkeys results in overgrooming of the fur, often causing an unattractive physical appearance in otherwise healthy monkeys. Should quality of the hair coat be considered as well, at least for animals on exhibit at zoos? Although these concerns may seem less significant than boredom in the individually housed monkey, there are other issues that are of much greater importance. Certain viruses, bacteria, and protozoa are endemic in primate colonies. Our ability to track and control the spread of a particular pathogen, whether it be salmonella, papilloma virus, or herpes viruses, will now be much more challenging when monkeys are routinely being rotated throughout the colony to establish new social groups. In the case of social housing in troops, dozens of monkeys may become infected with a simian immune deficiency virus following introduction of a single carrier. We must remember that historically the move toward individual housing was not to foster inhumane housing conditions on helpless animals, but rather to reduce the consequences of aggression and to minimize disease contagion. The proposed housing standards reflect the ascension of a new priority.

IMPLICATIONS FOR RESEARCH

In this section, we would like to provide some evidence for the concerns raised above by describing how social housing can influence the basic physiology of the nonhuman primate. We will offer several examples in the squirrel monkey and rhesus monkey that show the strong influence of social relationships on the endocrine, nervous, and immune systems. The studies clearly demonstrate that the selection of a particular housing condition will have a considerable influence on these physiological systems by determining the activity level at which the glands and cells function. Although the findings could be used as an argument for the need to house monkeys in social groups, they also serve to show that dominance relationships will introduce a new source of variability in physiological studies. In addition, we will review one physiological study that questions our ability to predict which conditions will reliably enhance the psychological well-being of nonhuman primates when the conditions are chosen by intuition alone. That is, we do not really know what the optimal housing conditions are for all of the exotic primate species used in research, especially at certain points in the life span (e.g., pregnancy, old age). In some species, the parturient female may actually prefer to be alone.

Let us begin by considering the possible influence of social housing on gonadal hormone release. Certainly, such an influence would be of paramount importance to a reproductive physiologist studying the pituitary-gonadal axis. Figure 1 illustrates the potentially strong effect of dominance rank on testosterone levels in the male monkey. The data were based on a 4-month assessment of testosterone levels in four male squirrel monkeys (*Saimiri boliviensis*) living in a stable social group. The cage environment would comply with current standards; it was 1.2 × 1.8 × 1.8 meters high and had 3 horizontal sitting perches. Food and water were provided ad libitum; fresh fruit and vitamin supplements were each provided twice per week. As can be seen, the onset of the annual increase in testosterone was markedly constrained by the social rank of the males at the start of the mating season. Only the two dominant animals showed the species-typical increase in testosterone and body weight; the third- and fourth-ranking males did not show the normal increase. Although this phenomenon is shown for only one group, we have demonstrated similar effects of psychological factors on hormone physiology in numerous other studies (Coe, Smith, & Levine, 1985).

A more dramatic illustration of this type of psychological influence on endocrine physiology can be seen in Figure 2. In a study of cortisol and testosterone

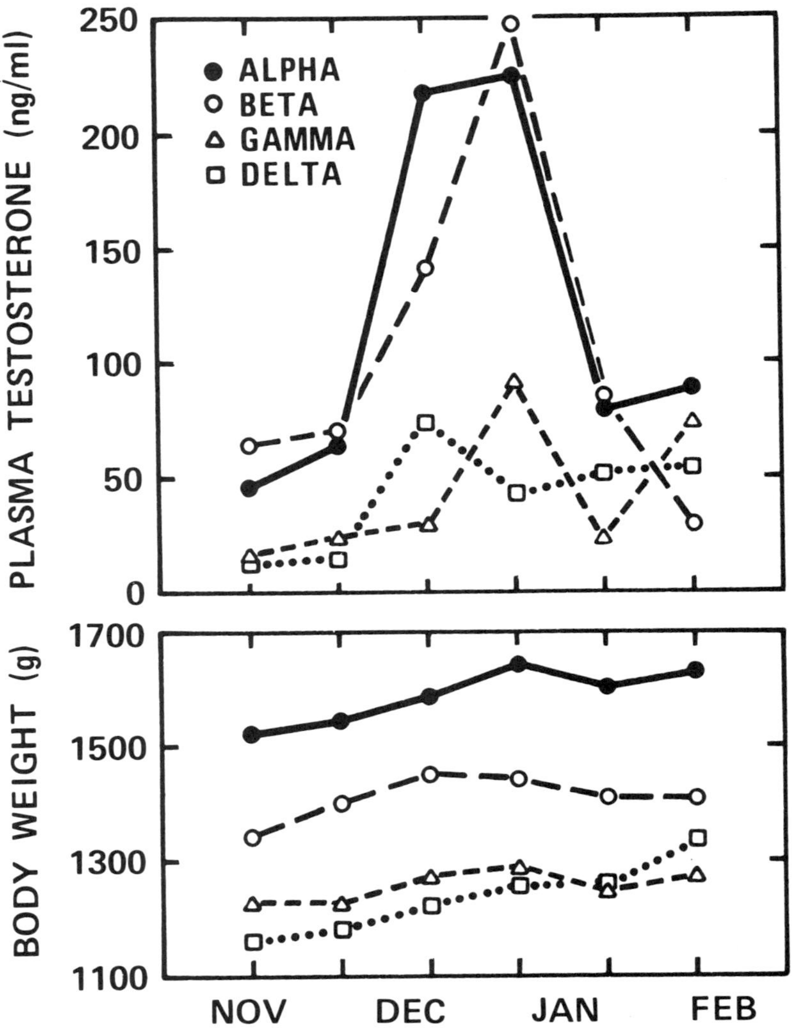

Figure 1 The influence of social rank on testosterone levels in a stable group of four male squirrel monkeys. Note the relative absence of seasonal hormone increases in the two low-ranking males.

levels in pair-housed male squirrel monkeys, we found that the hormone profile of dominant males was significantly different than that of their subordinate companions (Coe, Mendoza, & Levine, 1979). Hormone levels were determined at weekly intervals over a 5-week period, and the average testosterone levels are shown. In each of the eight pairs, the dominant male had higher testosterone levels. These differences in basal levels were also associated with a differential

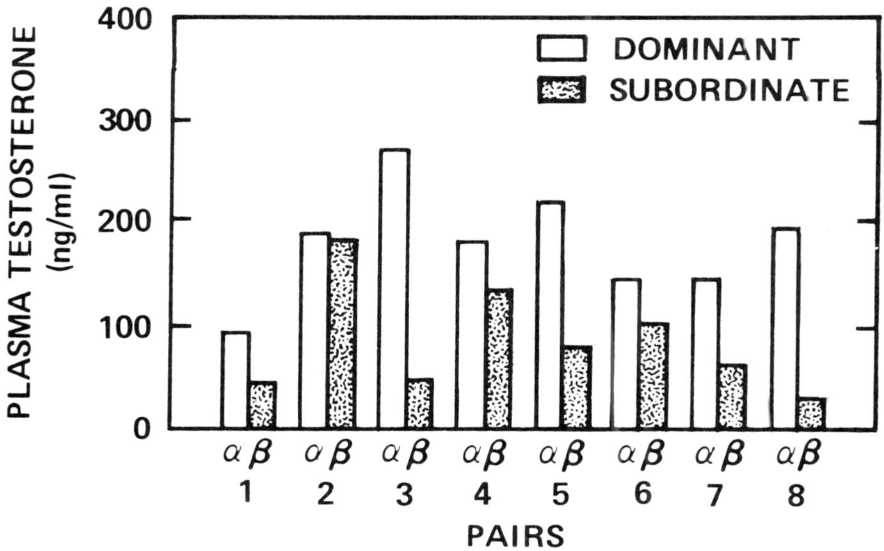

Figure 2 Differences in the testosterone levels found in dominant and subordinate squirrel monkeys housed in pairs. Modified from Coe, Mendoza, & Levine (1979).

hormone response to acute disturbance when we looked at the levels of testosterone 30 minutes after the first samples.

In addition, we have found that the formation of new social groups can have a strong effect on hormone levels for many weeks. Figure 3 portrays the results of a study of hormone levels in male squirrel monkeys before and after the formation of unisexual and heterosexual groups (Mendoza, Coe, Lowe, & Levine, 1979). Within 24 hours after the formation of three 3-male groups, one could delineate the influence of dominance rank on testosterone and cortisol levels in these monkeys. Further, the relative dominance rank of the males significantly influenced their behavioral and hormonal reaction to the introduction of five female monkeys at the end of the first month. Most importantly, only the dominant males showed an increase in gonadal hormone release in response to the female stimuli. [Note also that one subordinate male was separated from his previously nonaggressive male companions during the heterosexual phase in order to prevent fighting in this group, a problem of social housing mentioned earlier.]

This type of observation is not restricted to the squirrel monkey; nor is it applicable only to the pituitary-gonadal axis. In a recent longitudinal assessment of six pairs of male rhesus monkeys, we have found similar effects of dominance rank on testosterone and cortisol levels persisting over the period of a year. Moreover, we found that introduction of female subjects into the housing area had a significant effect on other physiological systems, including the immune

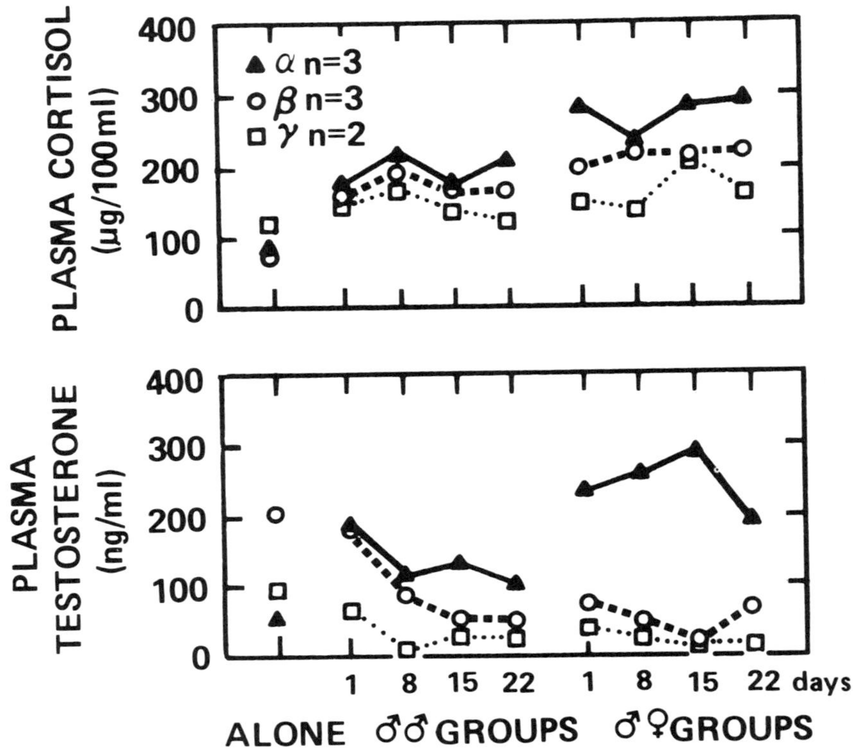

Figure 3 The effect of forming social groups on testosterone and cortisol levels in three 3-male groups of squirrel monkeys. Note that only the dominant males (triangle symbols) showed an increase in testosterone in response to the heterosexual condition. Reprinted from Mendoza, Coe, Lowe, & Levine (1979).

systems, of these male pairs. For example, both dominant and subordinate males showed a month-long decline in natural killer-cell activity following the psychosocial stimulation evoked by housing two females in cages adjacent to the cages of the male pairs. The types of behavioral processes that would mediate these physiological effects are illustrated in Figure 4. Here we can see the dramatic change in the affiliative and agonistic behavior of five pairs of bonnet macaque males following the presentation of a female monkey in a Plexiglas stimulus cage (Coe & Rosenblum, 1984). Males that had previously shown high levels of affiliative behavior no longer showed it and now exhibited aggressive behavior. We believe that these data underscore some of the logistical problems that managers of primate facilities will face in designing the optimal social housing environment for nonhuman primates.

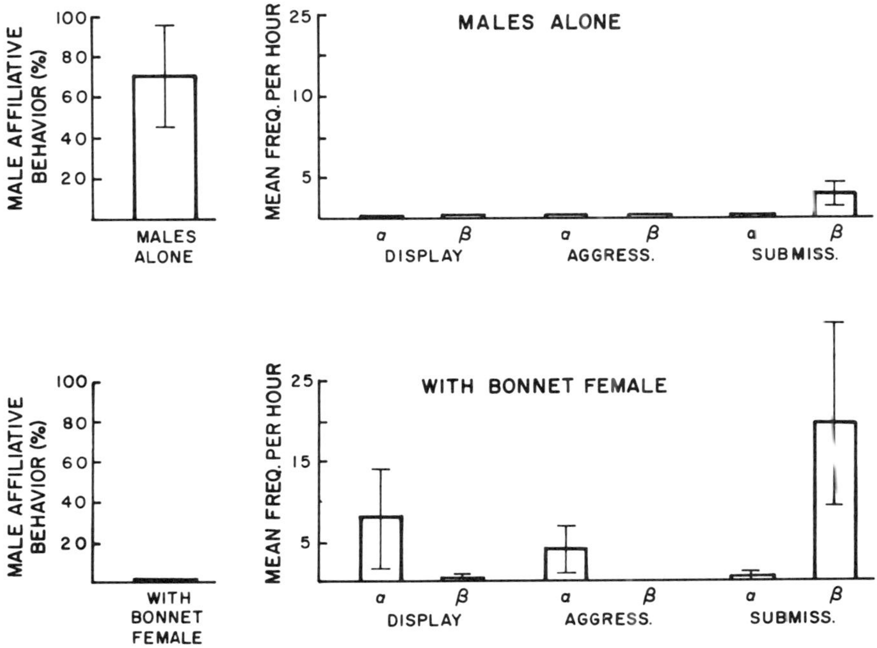

Figure 4 Changes in the positive social behavior of five pairs of bonnet macaque males induced by the proximity of a female conspecific. Reprinted from Coe & Rosenblum (1984). Mean (+/− S.D.) shown for five pairs in each condition.

Other studies have shown that social relationships and housing conditions can also influence brain neurochemistry, especially the monoamine systems. Measurements of monoamine metabolites in the cerebrospinal fluid of pair-housed squirrel monkeys have revealed that subordinate males show evidence of higher norepinephrine activity (Greene, 1989). In addition, studies on the vervet monkey have indicated that there is a strong effect of dominance rank on brain serotonergic activity (Raleigh et al., 1983). The psychobiological relationship is pervasive enough to determine the nature of the males' response to pharmacological agents that act through serotonin systems. That is, dominant male vervet monkeys reacted differently to certain drugs than did subordinate males.

Does this mean that we shouldn't house monkeys socially? No, it indicates that the decision to house monkeys socially must be integrated into the research design and be consonant with the research goals. For example, if we know that social housing can influence cardiovascular activity and the reaction to high-cholesterol diets, then we must be cognizant of this effect when we design a study and choose a particular housing condition for the research. In the individually housed squirrel monkey, blood-cholesterol levels are higher than in the socially housed monkey (Strickland & Clarkson, 1985). This fact may be an asset or a liability depending upon the goals of the research program.

The skeptical reader may still be unconvinced of the drawbacks to social housing. The studies cited so far have simply shown that social housing can influence basic physiological systems; they haven't really shown that social housing can have a negative effect on these systems. To address this issue, we will describe the results of a recent immunological study that we conducted at the University of Wisconsin (Coe, Ershler, Champoux, & Olson, submitted). Our collaborative research program has been concerned with the influence of psychological factors on immune responses in both young and old monkeys. We initially established that the rhesus monkey is an appropriate animal for studying immune senescence, the process of immunological decline at the end of the life span (Ershler et al., 1988). In a follow-up study, we were interested in determining whether immune responses would differ in old rhesus monkeys that were socially housed with juvenile monkeys, as compared to old monkeys that were housed either alone or with other old monkeys. This question had important practical ramifications, because we routinely socialize young monkeys by housing several of them with an old male and female monkey pair. To our eyes, the groups seemed to epitomize the essence of the new regulations on psychological well-being, at least for captive settings, and we embarked upon this study with the expectation that the old monkeys would show immune enhancement following social stimulation. That proved not to be the result, however, when we obtained the data back on lymphocyte proliferation and natural killer-cell responses. Contrary to our expectations, both old male and female rhesus monkeys showed lower responses on these two immune measures when compared to age-matched controls. Of course, with hindsight, one may be inclined to question whether this type of social group with 6–12 juveniles was not too stimulating for a monkey over 20 years of age. So a second experiment was conducted, presenting only one juvenile stimulus monkey to each of six old females using a repeated measures design. Again, a decrease in lymphocyte proliferation responses was observed over the 2-month social-housing phase.

These findings support the concern raised earlier in this paper about our ability to predict the outcome of all manipulations reliably, even when they are designed to be benevolent. To the lay person and to the trained eye, social housing in pens appeared to be an optimal living situation for the aged monkey. The old monkeys were more active and often engaged in extremely solicitous and positive interactions with the younger monkeys. Yet, we were misled by their overt behavior to assume that the physiological outcomes would also be positive. In fact, the higher activity levels induced by the complex social setting and presence of younger monkeys may have actually evoked the negative effect on immune responses by being too demanding. Old monkeys living alone are normally quite inactive, and the behavioral activity induced by social housing with juveniles, which we thought would be health-promoting, may actually have had the opposite effect. We come back to the question raised earlier in this paper: Have we conducted enough studies to be entirely sure of the best housing conditions for all primate species and for all age and sex classes? Since we probably have not, the regulations should provide for, and encourage, a fair degree of local autonomy and flexibility.

We would like to conclude this section by discussing two points that highlight important procedural issues that will affect many researchers if all monkeys must be socially housed. For certain types of research, the monkeys must be removed from their home cages and brought to a particular test or experimental area. This daily event is normally a routine procedure and may even be perceived as a positive event by the monkey—a break from the monotony of captive living. Certainly, this view is correct for some types of learning studies in which the monkey is able to perform tasks for food rewards. How will this animal-human interaction be affected by social housing? The answer is that we don't really know. For some species and certain tasks, the effect of social housing may be minimal; for others, the response to acute separation from a social partner could easily distort performance and invalidate years of baseline data collected previously. Will some monkeys become habituated to the regular separations from the partner during the test period, while others will not? Again, we would argue that a certain degree of local autonomy must be incorporated into the regulations in order to wisely implement the new requirements at the local level.

The final point illustrates the need for a thoughtful application of the social housing requirement. In this study, we evaluated whether weaning squirrel monkey infants affected the number and type of white blood cells found in their peripheral blood (Coe & Scheffler, 1989). The leukocyte pattern was used as a means of assessing how the monkeys were reacting to two routine procedures that are a regular aspect of colony management. First, the 6- to 12-month-old monkeys were weaned in order to breed the mothers, and the infants were then housed alone in another room for one week. At the end of the week, they were introduced into a juvenile peer group, and the leukocyte pattern was evaluated after 2 and 7 days to determine if the new social condition was beneficial or aversive. As can be seen in Figure 5, there was a predictable decline in lymphocytes (lymphocytopenia) and an increase in neutrophils (neutrophilia) in response to weaning and removal from the mother. With hindsight, the results from the subsequent phase may also seem predictable. The transition into an unfamiliar social setting during the second week was not perceived as a positive event and, for most of the 12 monkeys assessed, the juvenile peer group may have been even more disturbing than the individual housing. That is, social companionship is not necessarily positive unless the social partners are already familiar with one another. In other studies, we have found that the optimal way to wean a squirrel monkey is to separate several

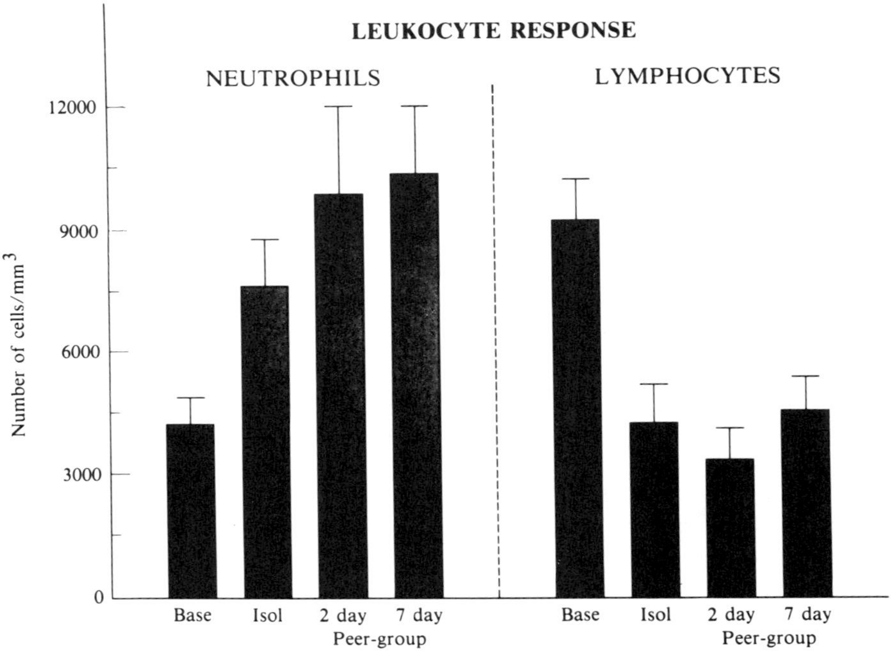

Figure 5 The effect of weaning on the number and type of white blood cells found in the peripheral blood of 12 infant squirrel monkeys. Note that the effect of rehousing into a peer group was even greater than that of isolating the weaned infant from its mother. Reprinted from Coe & Scheffler (1989).

familiar infants simultaneously by removing the mothers from the home cage environment. This is the only reliable way to minimize the immunosuppressive effects of weaning, at least as reflected by cell counts and the infant's ability to mount an antibody response to a virus (Coe, Rosenberg, Fischer, & Levine, 1987).

COST OF THE PROPOSED REGULATIONS

While some may dismiss the financial consequences of the new regulations as irrelevant to a discussion of the humane treatment of animals, it is important to be aware of the scope of the financial impact. Many of the cages currently available in American research institutions, even double-size ones that are currently used to house pairs of monkeys, may become illegal for pair-housing following acceptance of the new regulations. In particular, a requirement for change in vertical, as well as horizontal, dimensions will create a need for a cage design that is not currently used in most facilities (this requirement existed in the first set of proposed regulations, but has been revised in the second draft). Depending upon the materials and the specific configuration employed, the cost of the new cage will probably range from $2,000 to $4,000 per unit. Even a small facility with only 50 monkeys may be financially strapped when forced to generate this type of revenue at short notice. Larger primate centers with 500–1,500 monkeys will need significant infusions of new cash to meet the requirement for the larger cages.

The new cages will also have a number of hidden costs that will have an impact on the financing of primate facilities. The cost figure mentioned above did not take into consideration the fact that the cages will now be so large that they will probably have to be custom-made and assembled within each room. At this size and weight, the cages will probably not be mobile and will have to be cleaned within the room (thus, the cage design will have an effect on labor costs as well as hygiene). In some cases, the interior architecture of the room will have to be modified (e.g., the light fixtures, pipes, and ceilings may have to be raised). In addition, there will be a marked reduction in the housing capacity of existing animal quarters, which will seriously affect the cost-effectiveness of many research facilities.

The bottom line is that the new housing requirements will be extremely costly, and they come at a time when most research facilities are financially pressed to meet other regulations related to disease contagion. It is not clear where the necessary dollars will come from, and it should be emphasized that the regulations have been promulgated without a matching fund to help defray the incurred costs of modernizing American institutions to meet the revised standards. Obviously, most of the money will have to come from as yet unspecified state and federal sources. One could still argue that costs, at whatever level and consequence, should not be the driving factor in a discussion about the treatment of animals, but the cost issue must be considered. Indeed, should the funds become available, the new regulations could be a tremendous help to most primate facilities, allowing us to take a big step forward in our modernization plans.

GENERAL CONSIDERATIONS

Although this essay may seem too brief and the selected examples too few to question the overriding importance of the need for a change in primate housing, the cited studies are really representative of a much larger literature. We already know that the selection of a particular housing environment is critical for the successful conduct of a study. For some scientific goals, a social housing condition may be optimal. For others, the researcher may be able to incorporate the psychobiologic effects of group living into the research goals with more creative experimental designs. However, we must remain open to the possibility that individual housing may be the only way to carry out a particular study successfully. For example, in some research where the animal may have to be catheterized, it could be dangerous and inhumane to risk social housing. Similarly, in infectious disease research, contagion may be a problem that cannot be avoided without individual housing and quarantine.

We must also remain open to the possibility that the latest revision in housing standards may not be the final answer. For example, certain aspects of the cage-size requirements may still have some flaws in them. To illustrate this possibility, let us consider one possible housing problem: Should a particularly overweight rhesus monkey be housed in a larger cage or exercised more frequently than other rhesus monkeys?[1] According to some veterinarians, an obese monkey should be classified in a category set aside for species with much larger average body weights and, thus, would require a larger cage than a thin monkey of the same age and sex. Yet, Joseph Kemnitz (personal communication, 1989) of the Wisconsin Regional Primate Center has recently shown that obese rhesus monkeys are characterized by their relative inactivity, even prior to becoming overweight, and that they would probably move less frequently than their thinner counterparts in any size cage. If this proves to be true for obese primates of other species, then it suggests that overweight monkeys do not really need larger cages. In addition, the monkeys probably would not benefit from our attempts to exercise them against their wills. This is an extreme example, but it indicates the need for flexibility in the regulations as we begin to change the housing standards and requirements.

Our suggestions are not intended to undercut the new mandate for social housing of nonhuman primates, but to ensure that the regulations are appropriately formulated and implemented. To be assured that the new regulations are applied in a manner appropriate to local needs and priorities, we believe that the law should require that a general plan and goals for social housing of the maximum number of animals possible should be specified in the standard operating procedures (SOP) of each primate facility. A detailed SOP with local enforcement by the Institutional Animal Care and Use Committees (IACUC) would also serve to take some of the burden off federal inspectors who may not have the time during a brief inspection to evaluate all ramifications of a particular housing decision. In addition, during the unannounced USDA inspection, some pertinent investigators or staff may not be present to justify a particular housing decision. Clearly, a facility that has made little attempt to comply with the new regulations or one that is in obvious violation of its own SOP should be cited, and this decision must remain in the province of the USDA inspector.

Our other concern is that the new regulations be applied with an appreciation of the practical and financial ramifications for existing facilities. Although new facilities can be designed to meet the proposed codes, it may be extremely difficult for older facilities to comply with the new regulations in a short time frame, for

[1]It should be noted that the recent revision of the United States Department of Agriculture (USDA) regulations has taken this point into consideration. Body weight categories are based on species-typical weight, not on special conditions, such as obesity or pregnancy.

example, a period shorter than the annual budget periods and funding cycles of state and federal agencies. Obviously, the best possible solution would be for APHIS to work with local facilities to help them raise the awareness of administrators who may be somewhat reluctant to part with the funds needed to comply with the new requirements. Similarly, all of the relevant federal agencies and research institutions will have to convey to Congress the serious financial impact of the 1985 amendment with the hope that some new renovation funds will be forthcoming.

We will have to work together during this period of transition. Basic researchers will have to become more willing to interact regularly with veterinarians and administrators. For some researchers, the new regulations will require only a minimal change from current laboratory practices. In my own research with the small-bodied squirrel monkey, all of our monkeys are already housed in social groups and in large cages that exceed the new regulations. However, in many facilities, rhesus monkeys have not been traditionally housed in groups and pens, and the required changes and the resulting costs will be much greater. In anticipation of these new regulations, most facilities have already begun the process of complying, and in time, the new regulations will certainly have stimulated a marked improvement in the housing conditions for most nonhuman primates. But, given the enormous cost and the potential impact of the new regulations on the conduct of important biomedical research, it is hoped that we can implement some of these changes cautiously and with appropriate study as we proceed.

References

Coe, C. L., Ershler, W. B., Champoux, M., & Olson, J. *Psychological factors and immune senescence in the aged primate*. Manuscript submitted for publication.

Coe, C. L., Mendoza, S. P., & Levine, S. (1979). Social status constrains the hormonal response to stress in squirrel monkeys. *Physiology and Behavior*, *23*, 633–638.

Coe, C. L., Rosenberg, L. T., Fischer, M., & Levine S. (1987). Psychological factors capable of preventing the inhibition of antibody responses in separated infant monkeys. *Child Development*, *58*, 1420–1430.

Coe, C. L., & Rosenblum, L. A. (1984). Male dominance in the bonnet macaque: A malleable relationship. In P. Barchas, & S. P. Mendoza (Eds.), *Social cohesion: Essays toward a sociophysiological perspective* (pp. 31–64). Westport, CT: Greenwood Press.

Coe, C. L., & Scheffler, J. (1989). Utility of immune measures for evaluating psychological well-being in nonhuman primates. *Zoo Biology*, Suppl. 1, 89–100.

Coe, C. L., Smith, E. R., & Levine, S. (1985). The endocrine system of the squirrel monkey. In L. A. Rosenblum, & C. L. Coe (Eds.), *Handbook of squirrel monkey research* (pp. 191–218). New York: Plenum Press.

Ershler, W. B., Coe, C. L., Gravenstein, S., Schultz, K. T., Klopp, R. G., Meyer, M., & Houser, W. D. (1988). Aging and immunity in nonhuman primates: I. Effects of age and gender on cellular immune function in rhesus monkeys (*Macaca mulatta*). *American Journal of Primatology*, *15*, 181–188.

Greene, K. (1989). Unpublished Ph.D. dissertation, Stanford University Medical School.

Mendoza, S. P., Coe, C. L., Lowe, E. L., & Levine, S. (1979). Physiological response to group formation in adult squirrel monkeys. *Psychoneuroendocrinology*, *3*, 221–229.

Novak, M. A., & Suomi, S. J. (1988). Psychological well-being of primates in captivity. *American Psychologist*, *43*, 767–773.

Raleigh, M. J., Brammer, G. L., McGuire, M. T., Yuwiler, A., Geller, E., & Johnson, C. K. (1983). Social-status-related differences in the behavioral effects of drugs in vervet monkeys (*Cercopithecus aethiops sabaeus*). In H. D. Steklis and A. S. Kling (Eds.) (pp. 83–106) *Hormones, drugs and social behavior in primates*. New York: SP Medical.

Smuts, B. B., Cheney, D. L., Seyfarth, R. M., Wrangham, R. W., & Struhsaker, T. T. (1986). *Primate societies*. Chicago, IL: The University of Chicago Press.

Strickland, H. L., & Clarkson, T. B. (1985). Use of squirrel monkeys in cardiovascular research. In L. A. Rosenblum & C. L. Coe (Eds.), *Handbook of squirrel monkey research*, (pp. 295–314). New York: Plenum Press.

U.S. Congress, House. Improved standards for laboratory animals act (Public Law 99-148). (1985). *Congressional Record*, *131*, no. 175, H12335–12336.

CHAPTER 11

SURVIVAL AND REPRODUCTION AS MEASURES OF PSYCHOLOGICAL WELL-BEING IN COTTON-TOP TAMARINS (*SAGUINUS OEDIPUS*)

LORNA D. JOHNSON, ANDREW J. PETTO, AND PRABHAT K. SEHGAL

Psychological well-being is a poorly defined state that is difficult to measure in primates, including humans. In captive nonhuman primates, there have been few criteria for its provision, beyond the basic physiological and social needs of regular light cycles, limits of temperature and humidity, food and water, clean and spacious housing, and appropriate social groupings. There are even fewer criteria for its measurement. A recent review of the literature points to the difficulties in measuring well-being and warns of expensive and ineffectual changes that contribute more to the well-being of the researchers than to that of the animals (Petto, Novak, Fingold, & Walsh, 1989; Woolverton, Ator, Beardsley, & Carroll, 1989).

This research was supported by a grant from the National Institutes of Health, Division of Research Resources, RR00168. Animals used in this study were maintained in accordance with the guidelines of the Committee on Animals of the Harvard Medical School and those prepared by the Committee on the Care and Use of Laboratory Animals of the Institute of Laboratory Animals Resources, National Research Council (DHHS publication No. (NIH) 85–23, revised 1985).

This study was carried out under captive-bred wildlife registration PRT-699693 from the U.S. Fish and Wildlife Service to address specific health problems in *Saguinus Oedipus* that inhibit successful propagation and survival of the species, in accordance with 50 CFR 17.3(a) and CFR 17.21(g).

The growth of a colony through successful reproduction and survival of infants can be considered indicative of the overall well-being of the animals. These two endpoints can be measured without disturbing the animals and without costly experiments. This paper reports on a 12-year retrospective study of the effect of different techniques of rearing, cage sizes, and environment on reproduction and survival of captive cotton-top tamarins.

The cotton-top tamarin is an endangered species inhabiting a small area in northwest Colombia. It is estimated that there are more of these animals in captive colonies and zoos (Mittermeier 1986; Tardif, 1988) than in the wild (Neyman, 1977). The large colonies in research institutions provide the best opportunity for reconstitution of wild populations. Thus, it is important to be aware of those factors that promote increased production and survival.

Tamarins in captivity reach maturity at 18 to 24 months, are pair-bred, highly fertile, have a gestation period reported to be 170 to 184 days (Brand & Martin, 1983; Ziegler, Savage, Scheffler, & Snowdon, 1987), have twins or triplets more often than single births, and have no lactational anestrus, so that conception often occurs during the postpartum period. Thus, they are capable of having up to six infants annually. All captive colonies, however, have suffered from high infant mortality due to abuse or abandonment of the newborns.

MATERIALS AND METHODS

Colony Management

Cotton-top tamarins are housed indoors under conditions of constant temperature and humidity with 12-hour light cycles. Animals in the colony are fed a complex diet (Petto & Devin, 1988). Water with vitamin complex is offered ad libitum, and vitamin D3 is given weekly in marshmallows.

Rearing

Three methods of rearing the young have evolved over the years. Infants reared only by parents in the absence of other siblings are considered to be parent-reared. Parent-rearing occurred in the early years when juveniles were removed as twin pairs before the next litter was born and in later years when infants were removed if they were neglected or abused by the parents. When two or more litters are present with the parents, both the older siblings who took part in the rearing of the young and the younger siblings are considered to be family reared. Nursery-reared animals are those who are abused or neglected. They are hand-reared until weaning at approximately 3–4 months of age and then placed in large play cages with other nursery-reared animals until puberty.

From 1977 through 1981, a few family groups were established, and the parents and young lived in single cages measuring 84 cm H × 65 cm W × 69 cm D until the older siblings were approximately 18 months old. Then the older siblings were mated with other tamarins.

During the next 5 years (1982–1986), there were two changes in housing: Wooden nesting boxes replaced the stainless steel boxes, and upon the birth of each litter, an additional cage was connected by a plastic tunnel to provide double or triple volume for the families.

During the latter half of 1986, two rooms were set aside for some of the established families. In each room, eight cages measuring 1.5 m H × 1.2 m W × .76 m D were constructed 60 cm above the floor. The cages were identical except for their orientation in the room. Each cage contained wooden nesting boxes, perches, tree branches, and swings.

Database

The database of this study consists of 790 births at our Center from 1977 through 1988. The animal identification number, the origin of the animal (wild-caught, colony-born at the Primate Center, or received from another colony), date of birth and, where applicable, date and type of death, type of rearing, sex, the identification of the dam and sire, and the generation number by dam and sire were entered into the computer. Of 790 births, 628 (79.4%) were live-born. Nursery- and family-reared infants first appeared in substantial numbers in the years 1977–1981.

Statistical Analysis

The overall chi-square test for significance of differences in fertility and survivorship was computed for *r* and *c* contingency tables by standard methods (Everitt, 1977). When differences occurred, the category responsible was determined by analysis of residuals (Everitt, 1977) or partitioning chi-square (Kimball, 1954).

RESULTS

Rearing

Survivorship

In order to compare the 2-year survival rate (survival to adulthood) between the three different rearing groups, we selected the 449 infants born alive at 2 or more years before the end of the study: that is, between 1977 and 1986. Survival to reproductive age in the three groups is significantly different (Table 1). Exami-

Table 1

THE RELATIONSHIP OF REARING METHODS TO SURVIVAL IN 449 LIVE INFANTS BORN BETWEEN 1977 AND 1986

		Survival							
	No. of live born	7 Days		30 Days		1 Year		2 Years	
Rearing		**No.**	**%**	**No.**	**%**	**No.**	**%**	**No.**	**%**
Nursery	198	123	62.1	102	51.5	79	39.9	71	35.9
Parent	108	57	52.8	43	39.8	30	27.8	27	25.0
Family	143	132	92.3	123	86.0	115	80.4	110	76.9
Total	449	312	69.5	268	59.7	224	49.9	208	46.3

Note: Significance of differences determined by chi-square. Statistical test of homogeneity $\chi^2_{(2)} = 82.32$, $p < .05$.

nation of residuals shows that the difference is in the higher survivorship among infants raised in families. Survival past the first neonatal week, 3 months (the approximate time of weaning), and 1 year are also significantly different.

The crucial period leading to the ultimate survival after 2 years seems to be survival past the 1st week. At this time, survival past one week among nursery- and parent-raised infants is much lower (62% and 53% respectively) than among those raised in families (92%). After this first week, the differences in survival rates in the intervals from 7 days to 2 years are less variable, but there still appears to be a distinct advantage for family-reared tamarins. In these animals an additional 16% of the original cohort is lost at age 2 years, whereas the nursery- and parent-reared animals experience a further 26-28% loss of the original cohort before the age of two.

Fertility

There was no significant difference in the number of adult females who had given birth among those reared in the nursery, with parents, or in families (Table 2). There was also no overall difference in the fertility of the adult males in the three groups. However, partitioning of chi-square showed that the fertility rate among family-raised males was significantly higher than the other two groups. Colony-born adult females (70.6%) were significantly more fertile than males (54.5%). The difference between the sexes was due to the higher rate among the parent-raised females compared to comparably raised males.

Reproduction rates

The annual reproduction rates of colony-born dams and sires raised in the nursery, with parents, or in families are shown in Table 3. The three methods of rearing produced no significant difference in annual numbers of litters or live births among

Table 2

THE EFFECT OF REARING ON THE FERTILITY OF ADULT COLONY-BORN TAMARINS

	Females			Males		
	Total	Fertile		Total	Fertile	
Rearing	**no.**	**No.**	**%**	**no.**	**No.**	**%**
Nursery	37	22	59.5	34	15	44.1
Parent	19	16	84.2	8	3	37.5
Family	53	39	73.6	57	36	63.2
Total	109	77	70.6	99	54	54.5

Note: Differences significant as follows:
Males, $\chi^2_{(2)} = 4.13$. Partitioned: Family, $\chi^2_{(1)} = 4.02$, $p < .05$;
Parent & Nursery, $\chi^2_{(1)} = 0.11$; Male vs. female, $\chi^2_{(1)} = 5.77$, $p < .025$.

dams or sires. However, adjusted residuals showed that the rate of live births was highest among family-raised dams, lower among nursery-raised dams, and higher among parent-raised sires.

Housing

There was a significant difference between survival rates after 1 year according to whether infants were raised in family groups housed in single cages, double

Table 3

ANNUAL REPRODUCTIVE RATES OF COLONY-BORN DAMS AND SIRES RAISED IN THE NURSERY, WITH PARENTS, OR IN FAMILIES

	Nursery		Parent		Families	
Subject	**No.**	**Annual rate**	**No.**	**Annual rate**	**No.**	**Annual rate**
Dams	22		16		39	
Years Exposed	84		77		141	
Litters	69	0.82	74	0.96	138	0.98
Live Births	123	1.46	124	1.61	257	1.82
Sires	15		3		36	
Years Exposed	51		18		148	
Litters	43	0.84	22	1.22	125	0.84
Live Births	71	1.39	35	1.94	219	1.48

Note: Rate = mean number of live offspring per adult per year or the mean number of litters per adult per year.

Table 4

SURVIVAL OF INFANTS RAISED IN FAMILIES IN THREE DIFFERENT ENVIRONMENTS

Cages	Liveborn	Survival > 1 year
Single cages	36	27 (75.0)
Double cages	110	93 (84.5)
Family cages	24	15 (62.5)

Note: Test for homogeneity $\chi^2_{(2)} = 6.40$, $p < .05$
χ^2 Partitioned: Family $\chi^2_{(1)} = 1.51$, single + double $\chi^2_{(1)} = 4.89$, $p < .05$.

cages, or in family cages (Table 4). We compared survival after 1 year, because very few infants were born in the family cages at 2 or more years before the end of the study. Partitioning of chi-square showed the difference to be the lower *survival* in family cages.

Environment

Survival

A comparison of infant survival rates in the families before and after they moved into Rooms 1 and 2 is shown in Table 5. There was no significant difference in infant survival among those living in Room 1 before moving (93.0%) and after moving (81.8%), nor was there a difference between those living in Room 1 (93.0%) and Room 2 (83.3%) before entering either room. There was, however, a significant decline in infant survival to year 1 among those living in Room 2 (46.2%) after the families moved there.

Table 5

SURVIVAL AFTER 1 YEAR OF INFANTS BORN AND RAISED IN FAMILY GROUPS IN TWO ENVIRONMENTS

Environment	No. liveborn	Survival No.	Survival %
Room 1			
Before	43	40	93.0
After	11	9	81.8
Room 2			
Before	36	30	83.3
After	13	6	46.2

Note: Statistical test for homogeneity $\chi^2_{(3)} = 15.23$, $p < .005$. Room 2, before and after, $\chi^2_{(1)} = 6.77$, $p < .010$.

Table 6

ANNUAL REPRODUCTIVE RATES OF TAMARIN PARENTS LIVING IN FAMILY GROUPS, BEFORE AND AFTER ENTERING TWO DIFFERENT ENVIRONMENTS

	Room 1				Room 2			
	Before		After		Before		After	
	No.	Rate	No.	Rate	No.	Rate	No.	Rate
Months Exposed	444		197		500		207	
Litters	50	1.35	16	0.97	54	1.30	22	1.28
Live Births	97	2.62	31	1.88	93	2.23	45	2.61

Reproduction

There was no statistically significant difference in the annual reproduction rates (number of litters/years exposed), or live-birth rates among families whether measured before or after these families moved into either Room 1 or 2 (Table 6).

DISCUSSION

The social environment of primates early in life has been shown to have profound effects upon their social behavior as adults (Locke-Haydon & Chalmers, 1983; Pryce, 1988). Several studies have shown that survival among captive cotton-top tamarins is improved by living in extended families (Kirkwood, Epstein, & Terlecki, 1983; Snowdon, Savage, & McConnell, 1985; Tardif, Carson, & Clapp, 1986). This social grouping most nearly approximates groups observed in the wild, which consist of small polyandrous groups of adult females with their offspring (Sussman & Kinzey, 1984). In captivity, the adult pair is monogamous, each male mating with only one fertile female. Consider the vast difference between the wild environment and that of being reared with parents alone or being hand-raised in a nursery. There have been no comparisons of the effects of these captive environments on the survival of infant tamarins or on their reproductive performance as adults.

Our data show that rearing captive cotton-top infants in family groups significantly increased their rate of survival to adulthood (77%). By comparison, only 25% of those reared alone with parents and 36% of those reared in the nursery survived to adulthood.

Different methods of rearing females had no significant effect on their fertility, and among males the overall test for association between rearing and fertility was not significant at the .05 level. However, partitioning of chi-square showed that family-reared males were significantly more fertile than males in the other two groups combined. The fertility rate for colony-born animals was significantly higher for females than males. This was due to the significant difference in fertility between parent-raised females (84%) and parent-raised males (38%).

The data show that family rearing is highly desirable in terms of colony management. The combination of high survival, fertility, and reproductive rates produces a larger number of offspring in a given time period. Several studies have shown that family groups in large cages and complex environments have high infant-survival rates (Clapp & Tardif, 1985; Snowdon et al., 1985). Thus, our data (Table 4) on the effect of cage size and environment on the survival of infant tamarins living in family groups were unexpected. There was no significant difference in infant survival in single cages (75%) or double cages (85%) with a low order of environmental complexity and in room one with larger family cages (82%) with a high order of complexity.

Of particular interest was the lower survival rate in one of two rooms in which large family cages were identical except in their orientation in the room and to each other. The family cages in Rooms 1 and 2 were the same size, made of the same materials and had the same accoutrements. The orientation of the cages, however, produced a different environment. In Room 1, four cages were arranged along opposite walls facing each other. The front of the cage was the longer side and was made of wire grid through which the families could see conspecifics in similar cages and on which they could climb. The short side was made of laminate and had perches attached. It separated adjacent cages so that no visual or physical contact with immediate neighbors was possible. In Room 2, the eight cages were lined on one wall and faced the blank, opposite wall. The front of the cage was the short side made of grid, the long laminate side between cages had perches attached. Thus, the families' view of the world was more limited, they could not see conspecifics, and there was less area on which to climb and more perch area. The rooms were identical except for the wall coloring: Room 1 was painted a medium blue and Room 2 was pale green.

The low survival rate occurred in cages where the families had less climbing room, more perching room, and no visual contact with conspecifics (Room 2). Several investigators have observed signs of stress and aggression (Kirkwood, Epstein, & Terlecki 1983; French & Snowdon, 1981), sometimes resulting in abortion or decreased survival of infants and fighting among adults when family groups were exposed to conspecifics (Epple, 1970). Therefore, animals in Room 2 were expected to have the highest rate of infant survival. These same families had less perching and climbing room in the cages they occupied before entering this room. Therefore, it is difficult to assign either climbing and perching space or amount of visual contact to the decrease in infant survival. Further studies are needed to determine the cause of the low survival.

If infant survival is a correlative of well-being, these data show that subtle changes in environment can make an unexpected difference. This finding emphasizes the difficulty in applying simple models to provide environments for the well-being of captive animals and cautions us to test new designs on a small scale before investing in costly, large-scale changes.

References

Brand, H. M. (1981). Husbandry and breeding of a newly established colony of cotton-topped tamarins (*Saguinus oedipus oedipus*). *Laboratory Animals*, *15*, 7–11.

Brand, H. M., & Martin, R. D. (1983). The relationship between female urinary estrogen excretion and mating behavior in cotton-topped tamarins, *Saguinus oedipus oedipus*. *International Journal of Primatology*, *4*, 275–290.

Clapp, N. K., & Tardif, S. D. (1985). Marmoset husbandry and nutrition. *Digestive Diseases and Sciences*, *30*, 17s–23s.

Epple, G. (1970). Maintenance, breeding, and development of marmoset monkeys (Callithricidae) in captivity. *Folia Primatologica*, *12*, 56–76.

Everitt, B. S. (1977). *The analysis of contingency tables*. London: Chapman and Hall.

French, J. A., & Snowdon, C. T. (1981). Sexual dimorphism in responses to unfamiliar intruders in the tamarin, *Saguinus oedipus*. *Animal Behaviour*, *29*, 822–829.

Kimball, A. W. (1954). Short-cut formulae for the exact partition of χ^2 in contingency tables. *Biometrics*, *10*, 452–458.

Kirkwood, J. K., Epstein, M. A., & Terlecki, A. J. (1983). Factors influencing population growth of a colony of cotton-top tamarins. *Laboratory Animals*, *17*, 35–41.

Locke-Haydon, J., & Chalmers, N. R. (1983). The development of infant-caregiver relationships in captive common marmosets (*Callithrix jacchus*). *International Journal of Primatology*, *4*, 63–81.

Mittermeier, R. A. (1986). Primate conservation priorities in the neotropical region. In K. Benirschke (Ed.), *Primates : The road to self-sustaining populations*, (pp. 221–240). New York: Springer-Verlag.

Neyman, P. P. (1977). Aspects of the ecology and social organization of free-ranging cotton-top tamarins *Saguinus oedipus* and the conservation status of the species. In D.G. Kleiman (Ed.), *The biology and conservation of Callitrichidae*, (pp. 39–71). Washington, DC: Smithsonian Institution Press.

Petto, A. J., & Devin, M. (1988). Food choices in captive common marmosets (*Callithrix jacchus*). *Laboratory Primate Newsletter*, *27*, 7–9.

Petto, A. J., Novak, M. A., Fingold, S. A., & Walsh, A. C. (1989). The search for psychological well-being in nonhuman primates: Information resources. *Science and Technology Libraries*, *10*, 101–127.

Pryce, C. R. (1988). Individual and group effects on early caregiver-infant relationships in red-bellied tamarin monkeys. *Animal Behaviour*, *36*, 1455–1464.

Snowdon, C. T., Savage, A., & McConnell, P. B. (1985). A breeding colony of cotton-top tamarins (*Saguinus oedipus*). *Laboratory Animal Science*, *35*, 477–480.

Sussman, R. W., & Kinzey, W. G. (1984). The ecological role of the Callithricidae: A review. *American Journal of Physical Anthropology*, *64*, 419–449.

Tardif, S. D., & Colley, R. (1988). *International cotton-top tamarin stud book* (2nd ed.). Oak Ridge, TN: Oak Ridge Assoc. Universities.

Tardif, S. D., Richter, C., & Carson, R. L. (1984). Effects of sibling-rearing experience on future reproductive success in two species of Callithricidae. *American Journal of Primatology*, *6*, 377–380.

Tardif, S. D., Carson, R. L., & Clapp, N. K. (1986). Breeding performance of captive-born cotton-top tamarin (*Saguinus oedipus*) females: Proposed explanations for colony differences. *American Journal of Primatology*, *11*, 271–275.

Woolverton, W. L., Ator, N. A., Beardsley, P. M., & Carroll, M. E. (1989). Effects of environmental conditions on the psychological well-being of primates: A review of the literature. *Life Sciences*, *44*, 901–917.

Ziegler, T. E., Savage, A., Scheffler, G., & Snowdon, C. T. (1987). The endocrinology of puberty and reproductive functioning in female cotton-top tamarins (*Saguinus oedipus*) under varying social conditions. *Biology of Reproduction*, *37*, 618–627.

CHAPTER 12

NATURALISTIC ENVIRONMENTS AND PSYCHOLOGICAL WELL-BEING

CHARLES T. SNOWDON

PSYCHOLOGICAL WELL-BEING AND CONTROL OF ENVIRONMENT

The development of a criterion for psychological well-being in human beings has been extremely difficult (Diener, 1984), so the development of clear criteria for psychological well-being in nonverbal organisms is going to be even more difficult. Well-being is defined in the dictionary as "happiness, comfort, welfare"; but these terms are equally difficult to define clearly. Frequently psychological well-being is defined as the absence of stress, but, as Novak and Suomi (1988) and others in this volume have noted, the absence of stress is not necessarily good. We humans seek moderate levels of stress for ourselves in our work and our recreation, and we appear to prefer a moderate level of challenge to either a totally stress-free environment or a highly stressful environment. A moderate level of stress that we can regulate ourselves appears to be preferable. Given that human beings seek constant intellectual, physical, and social challenges, one might argue that psychological well-being could be defined in terms of our ability to adapt to changes in our environments.

Preparation of the manuscript and the author's research were supported by a Research Scientist Award, MH 000,177 and USPHS Research Grants MH 29,775 and MH 35,215. I am grateful to Sheryl Hayes, Patricia McConnell, Anne Savage, Tina Widowski, and Toni Ziegler for helping to develop our approach to naturalistic housing and for discussion of the ideas in this chapter.

Some recent studies on the rhesus macaques (*Macaca mulatta*) indicate that control over the environment is an important component of psychological well-being and of the ability to adapt to environmental change. Hanson, Larson, and Snowdon (1976) presented adult monkeys with loud, continuous white noise. Monkeys in one group could terminate the noise by pressing a lever. Monkeys in a second group were yoked to the first group, experiencing exactly the same amount of noise stimulation, but they had no lever to terminate noise. A control group received no noise stimulation. Subsequently, the lever that controlled noise termination was removed from the monkeys that had previously had control. During the experiment, both cortisol levels and social behavior after noise exposure were monitored. The results were quite striking. The group with control over noise termination had low cortisol levels that did not differ from those of the control monkeys. The monkeys with no control had significantly elevated cortisol levels. When control over the noise was removed, the cortisol levels rose to even higher levels than those of monkeys that had never had control. Monkeys with control over noise showed low levels of aggression, those without control over noise showed intermediate levels, and those that had lost control over noise showed the highest levels of aggression.

Mineka, Gunnar, and Champoux (1986) established peer groups of rhesus monkey infants under two conditions. In one condition, monkeys could press bars, pull chains, and engage in other operant activities for small treats of food. Animals in the other condition received the same food rewards at the same time, but the delivery of rewards was unrelated to the manipulation of the environment. This relatively minor manipulation had important effects on subsequent behavior. The monkeys that had to work for food showed less fearfulness in response to threatening stimuli and showed better coping responses when separated from other monkeys than did the monkeys that received food without working for it. There was a tendency for the working monkeys to have lower cortisol levels than the yoked control group. Mineka et al. (1986) discussed their results in terms of control over the environment. The animals with previous experience of controlling their environment might be said to have developed a confidence that they could control their environment in novel situations.

The results of both the Hanson et al. (1976) and Mineka et al. (1986) studies suggest that providing monkeys with a sense of control over their environments can reduce stress levels, increase coping skills in novel situations, and reduce levels of aggression. The psychological well-being of captive primates would benefit from providing them opportunities to gain control over their environment, and I will argue that these opportunities are more readily provided in naturalistic enclosures where animals can interact more with the environment. However, the results of the Hanson et al. (1976) study caution that once animals have been provided with a sense of environmental control, the removal of this control can be harmful.

CRITERIA FOR SUCCESSFUL CAPTIVE ENVIRONMENTS

For the past several years, my research has involved working with a captive colony of an endangered tamarin species. Two major goals have influenced our approach to research and to husbandry. First, we must maintain a colony that is healthy, that is reproducing and rearing its own offspring, and that retains as much normal social behavior as possible, since opportunities to study these animals in the wild are becoming increasingly rare. Second, it is essential to maintain an active research program that does not compromise the reproductive success of the colony. An active research program with endangered primate species is important for two reasons. We must learn as much as possible about these animals before they become extinct, and we can justify the maintenance of our colony through an active research program.

Our goals parallel those of the Duke Primate Center (see M. K. Izard, chapter 15 in this volume), and to meet these goals, we have developed large naturalistic enclosures for our animals, which are equipped with complex networks of ropes, boards, and branches to simulate an arboreal environment. The development of these naturalistic environments for captive tamarins is important to their well-being for three reasons. First, the strictest definition of a successful captive environment for endangered primates can best be met in conditions where key features of a species' natural social and physical environment are provided (Snowdon, 1989). Second, many of the research goals of primate behaviorists are best met by studying animals in naturalistic environments. Third, housing tamarins in more naturalistic environments has compelled us to develop less invasive ways of gathering research information, which in turn can provide for better psychological well-being. In describing the criteria for evaluating naturalistic environments, I will focus on each of these three points.

Veterinary Criterion

I have developed (Snowdon, 1989; Snowdon & Savage, 1989) a three-staged criterion for the evaluation of captive environments for endangered primates. The first stage is the veterinary criterion, which is a close approximation to the current regulations of the National Institutes of Health and the United States Department of Agriculture (USDA). Animals must be provided with sanitary environments, with clean and nutritionally adequate food, with sufficient air changes, and with stable temperature and lighting conditions. The goal of the veterinary criterion is to provide animals that are in good physical and nutritional health and free from disease.

Biological Criterion

The second stage is the biological criterion, based on the Darwinian notion of fitness as measured through reproductive success. This second stage states that an effective captive environment is one in which animals can reproduce successfully. Most primate researchers in this country have accepted this criterion over the past two decades. The natural habitats of most species of primates are severely threatened, and it is irresponsible to be dependent upon wild-caught primates to maintain captive populations.

However, two definitions of reproductive success are in use. For some, the female delivering infants is a sufficient criterion whether or not that female and her social group are capable of rearing the infants successfully. Some places are satisfied to hand-rear primates, but hand-rearing of infants is expensive and time consuming compared with the cost of natural rearing, and the hand-reared animals are apt to be deficient in parental skills when they become reproductively mature (see L. Johnson, A. Petto, and P. K. Sehgal, chapter 11 in this volume).

Reproductive success is better measured by parents caring for their own offspring. In the family Callitrichidae, the marmosets and tamarins, there is an extensive literature indicating that young animals must have hands-on experience in caring for someone else's infants in order to acquire parental care skills that they need later on (Epple, 1978; Hoage, 1977; Snowdon, Savage, & McConnell, 1985; Tardif, Richter, & Carson, 1984a,b). Marmosets and tamarins without prior experience with infants react to infants on their backs by pushing them off and biting them. Females without prior experience do not position infants properly for nursing. Such behavior is not conducive to successful infant rearing.

Cleveland and Snowdon (1984) described the interactions of other group members with infant cotton-top tamarins (*Saguinus oedipus*) and found that 7-month-olds initiated play with infants but were not involved in carrying them. Only with the second set of infants did older siblings (10 months and older) develop the appropriate caretaking behavior. If monkeys must have direct experience with two successive sets of infants in order to acquire parental competence, then they must be housed in captive environments large enough to accommodate a relatively large number of animals. Snowdon et al. (1985) surveyed the published reports of colony success with cotton-top tamarins and found a correlation between cage size and complexity and reproductive success. The most instructive data come from the Marmoset Research Facility at Oak Ridge, Tennessee. Gengozian, Batson, and Smith (1978) reported an infant survival rate of 25% using the minimum-sized cages approved by the USDA. In the late 1970s and the 1980s, breeding animals were housed in large family-sized cages, and the reproductive success has increased significantly (Tardif et al., 1984a,b).

Behavioral–Ecological Criterion

The third stage is the behavioral-ecological criterion, which is based on the rationale used by those of us who breed endangered species. One goal, probably overly idealistic, is that we can maintain a species through captive breeding until the habitat is rehabilitated, and animals can be released into the wild. Although there are logical and practical problems with reintroduction of captive-born animals into the wild (Brambell, 1977; Konstant & Mittermeier, 1982), there has been recent success in reintroduction of golden lion tamarins (*Leontopithecus rosalia*) to their habitat in Brazil (Kleiman et al.,1986). The behavioral-ecological stage suggests that a good captive environment is one that maintains in a captive population the motoric, social, cognitive, and other skills that would be needed by animals were they to be reintroduced to the wild.

There are several consequences of considering the behavioral-ecological stage as part of a criterion of psychological well-being. First, each species must be treated separately. No general set of rules can be written that will describe adequate psychological well-being for all primate species. Each species will need its own requirements for an adequate captive environment. Second, we will need to have a good knowledge of the normal behavior of a species in its natural environment; we must have information on the ontogenetic mechanisms underlying behavior. If many of the important skills of an animal are acquired through observational learning and social reinforcement, then such opportunities must be provided in captivity. Third, we need to provide appropriate physical environments that allow expression of species-typical behavior. Arboreal animals must have structures where they can live and feed well above the ground or floor. Finally, animals need to develop a sense of control over their environment so that they can deal with unpredictable environmental changes.

Providing an analogue to the normal physical environment can be important in developing sensorimotor skills. Arboreal monkeys, for example, need to develop careful eye-limb coordination to move in a three-dimensional environment. They must be able to judge distances and evaluate how much "give" to expect when leaping from one branch to another. Because direct pathways between two points are rare in a forest, animals must learn detour behavior. Clawed arboreal primates, such as marmosets and tamarins and many of the prosimians, will need a variety of soft structures such as branches, boards, and ropes, which they can cling to with their claws while moving. Hard, impervious materials such as stainless steel are not only inappropriate but potentially dangerous. Thus, a captive environment for clawed arboreal primates should have a complex of elevated structures made of wood and fiber, with some structures fixed and unmoving and others flexible. Different diameters of materials will mimic the variety to be found in natural environments. Different species might need different types of configurations (see M. K. Izard, chapter 15 in this volume).

These types of environment can be changed relatively frequently, both to provide a means of sanitizing the environment and to provide novelty for the monkeys in terms of new configurations. Such environments work. In our cotton-top tamarin colony, we have never had an animal become ill or die of anything related to the use of wood or fiber in the cages. We have also completed a pilot study where we released captive-born monkeys to a rain forest at Monkey Jungle in Florida, where three of the four monkeys showed excellent locomotor skills, being able to find their way through the canopy, to jump across gaps between branches, and to move on branches and trunks of widely varying diameters. Our captive environments come close to providing the locomotory skills needed in the wild.

Another area where naturalistic captive environments is important is social learning. Many biologists often assume that there are instincts for parental care, for predator avoidance, for social communication, and for foraging. But marmosets and tamarins need to have direct exposure to infants in order to acquire parental care skills. There is no innate parental care system.

Some monkey species must learn predator recognition skills. Mineka and her colleagues (Mineka, Davidson, Cook, & Keir, 1984; Mineka, Keir, & Price, 1980) have shown that rhesus monkeys must learn to be afraid of snakes, but that they can learn this fear through observation. More recent work (Mineka & Cook, 1988) shows that rhesus monkeys can learn snake fears by watching videotapes of wild-caught monkeys reacting to a snake. This result suggests that a library of video training tapes might be developed to preserve the behavior of the remaining wild monkeys for tutoring captive monkeys.

We have tested cotton-top tamarins with snakes (Hayes & Snowdon, 1990) and found that captive-born tamarins never exposed to snakes showed few fear responses, which habituate readily. However, tamarins reared in environments where they might have been exposed to snakes showed stronger fear responses that did not habituate. If one were to reintroduce tamarins to the wild, it would be essential to determine if they needed training in recognizing predators.

The development of appropriate form and usage of vocalizations appears to involve observational learning and social reinforcement. Although some aspects of vocal behavior (especially in squirrel monkeys, *Saimiri sciureus*) appear to be genetically determined (Herzog & Hopf, 1984; Newman & Symmes, 1982), studies of other species indicate that both appropriate vocal structure and appropriate usage depend upon social learning. Seyfarth and Cheney (1986) showed that young vervet monkeys (*Cercopithecus aethiops*) depend upon adult modeling and reinforcement in the development of both the predator alarm system and the grunt vocalization system.

A good captive environment should also promote appropriate foraging skills. In the reintroduction program with golden lion tamarins (Kleiman et al., 1986), extensive training was done to prepare tamarins to find food that was distributed over time and space and that was frequently cryptic or hidden. Tamarins were required to open bird's eggs, to handle whole fruits rather than cut-up fruit, and to extract food from the natural environment. McGrew, Brennan, and Russell (1986) described a sap feeder for use with captive common marmosets (*Callithrix jacchus*) and pygmy marmosets (*Cebuella pygmaea*). In the wild, both species excavate holes in trees to extract sap. Foraging tasks are beneficial in environmental enrichment (Chamove, Anderson, Morgan-Jones, & Jones, 1982; Plimpton, Swartz, & Rosenblum, 1981; Rosenblum & Smiley, 1984). Each of these tasks functions not only to develop specific foraging skills but to provide control over the environment.

In working with primates, it is more conservative to assume that behavioral competence arises through social and environmental interaction rather than through maturational and genetic variables. This means that naturalistic environments are important in providing young primates with the social and physical environments needed to acquire adult behavioral competence. A naturalistic environment can improve psychological well-being by increasing the likelihood that animals will develop perceptions of control over their environment without the need for complex operant tasks. The experiences of finding food located in different parts of the environment, of adapting locomotor patterns to changing cage structures, of adapting to the complex social relationships provided by a natural social environment can lead to less stressful, less aggressive animals that have better skills for coping with environmental change.

NATURALISTIC ENVIRONMENTS AND THE GOALS OF BEHAVIORAL RESEARCH

There are three goals for many of us doing behavioral research on captive primates, and each of these goals requires some degree of naturalistic housing for animals. First, the data gathered on captive animals can be used to supplement and complement observations made in the field (Miller, 1977). However, if data gathered on captive animals are to be useful in completing or expanding upon data gathered in the field, then relatively natural environments or relatively natural social groupings must be used.

Second, the behavior of nonhuman primates is frequently used to provide insights about the origins or mechanisms of phenomena of human behavior. If the parallels between nonhuman primate behavior and human behavior are to be valid, then one must have an understanding of the natural behavior of each species.

Finally, a major benefit of research on nonhuman primates is to illustrate differences between species. How do different organisms respond to similar problems? If they respond in different ways, then we can learn much from examination of the evolutionary, ecological, social, or ontogenetic differences between these problems or species. The diversity of responses that other species display to problems can provide novel insights into human behavior.

There are many possible examples to illustrate how research on captive animals complements field studies. I will give a few examples based on research with tamarins. Gautier and Gautier-Hion (1982) noted that the study of primate vocal communication in the wild is frustrated by the inability of observers to get close enough to monkeys to record intragroup and contact vocalizations. Thus, field studies are biased toward long-distance signals. In our captive colony of cotton-top tamarins, we have been able to obtain more detailed descriptions of vocal structure and function (Cleveland & Snowdon, 1982) than we have been able to obtain in the wild (A. Savage & C. T. Snowdon, personal observations).

The only field study on cotton-top tamarins (Neyman, 1977) described some aspects of territorial encounters. However, McConnell and Snowdon (1986) simulated territorial encounters between captive groups by increasing the amplitude of calls of an unfamiliar group, and thereby obtained detailed descriptions of the vocal and behavioral actions of territorial encounters that Neyman could not.

The issue of how and when the young disperse to new groups and how reproductive exclusivity is maintained in tamarins is an important question that could only be answered in the field through years of observation of marked populations. However, several laboratory analogues can provide information on these points. French and Snowdon (1981) found a sexual dimorphism in how cotton-top tamarins respond to intruders. Females scent-marked extensively to intruders of both sexes but showed little threat behavior. Males showed extreme threat behavior toward intruding males but not to females.

We subsequently found that some component of the scent of a reproductive female serves to block ovulation in subordinate females in a group (Savage, Ziegler, & Snowdon, 1988) and that stimulation by a novel male is also critical to induce ovulation (Ziegler, Savage, Scheffler, & Snowdon, 1987). These results suggest that in the wild, males should remain in their natal groups and attack intruding males. Females would have more flexibility in dispersal, since they would be reproductively suppressed by the dominant female.

In a clever captive study, McGrew and McLuckie (1986) attached plastic ducting to cages of cotton-top tamarins and ran the tubing through the hallway to empty cages. Young, reproductively suppressed females were the first to enter the ducting, and they traveled greater distances and were more likely to move to new cages, suggesting that in the wild, females may be the sex that disperses.

In a closely related species, the golden lion tamarin, French and his colleagues have obtained quite different results. In an intruder study with golden lion tamarins, French and Inglett (1989) found that female golden lion tamarins reacted with severe aggression toward female intruders, while males were relatively calm in the presence of intruders. Instead of the complete suppression of ovulation of subordinate females that we have found with cotton-top tamarins (French, Abbott, & Snowdon, 1984; Ziegler et al., 1987), golden lion tamarin mothers and daughters appear to ovulate in synchrony (French & Stribley, 1987). Thus, an intruding female lion tamarin might be a greater threat to the reproductive female than an intruding cotton-top tamarin would be to the reproductive female. These results lead to predictions about species differences in dispersal that could be examined in future field studies.

These examples illustrate how simulations of naturalistic environments for captive animals can be used to learn more about the natural lives of animals than might be possible in nature. They also illustrate the importance of attending to behavioral and physiological differences, even between closely related species.

NATURALISTIC ENVIRONMENTS AND NONINVASIVE RESEARCH TECHNIQUES

To both the general public and much of the research community, research is often synonymous with invasiveness. It is thought that animals must be kept isolated from one another in carefully controlled environments. One consequence of housing animals in more naturalistic environments is that alternative research methods can be used to minimize or eliminate invasiveness. We have developed noninvasive research methods with our colony. More often than not our data can be gathered with greater efficiency than is possible using traditional methods, and we often obtain data that are of higher quality, because the methods of data collection no longer interact with the data itself. Naturalistic environments do not need to impede research, and the reduction of invasiveness leads to increased well-being.

I will illustrate this argument with four examples from our research with cotton-top tamarins. Because this is an endangered species, we have tried to avoid any invasive techniques. However, the methods to be described can easily be used with a variety of species, whether endangered or not.

Comparative psychologists have been concerned with evaluating the learning ability or cognitive skills of monkeys. Miles and Meyer (1956) used the Wisconsin General Test Apparatus (WGTA) to study learning-set acquisition in common marmosets. The marmosets showed no improvement in learning-set performance until after 900 problem sets. In contrast, Jeff Lande (unpublished observations) studied learning-set performance in our cotton-top tamarins using a portable WGTA placed at the side of the animals' home cages. Lande found 90% performance within 32 learning-set problems—comparable to the best data reported by Harlow (1949) for well-habituated rhesus monkeys. Subsequently, we have used this apparatus to study color discrimination in cotton-top tamarins (Savage, Dronzek, & Snowdon, 1987). Individual subjects received more than 3,000 trials and would work for 35–45 minutes per day. Testing in the home cage eliminates the need to separate animals from their social group and eliminates the need for the usual lengthy habituation to the novel testing environment.

We have used playbacks of sounds through speakers in the home cage to study auditory discrimination. Cleveland and Snowdon (1982) described eight forms of brief (<100 ms), chirp-like vocalizations that were used in highly distinctive contexts. Bauers and Snowdon (1990) selected the two chirp forms that were most similar in acoustic structure, differing only in onset frequency and frequency range, and played them back to tamarins, using spontaneous behavioral and vocal responses as an index of discrimination. In the absence of any other contextual cues, the two stimuli played back through the speakers elicited different responses, indicating that the animals could discriminate between these two calls on the basis of acoustic information alone. Furthermore, a significant discrimination appeared on the very first trial of testing. Similar acoustic discrimination studies using operant methods appear to have required on the order of 30,000 trials to establish a discrimination (Petersen et al., 1984). The use of playback techniques to test animals in their home environments is considerably more efficient.

Our interest in managing our colony more effectively and learning more about the reproductive biology in cotton-top tamarins has led to the development of new techniques for monitoring hormonal status. Traditionally, serum measurements have been used for hormonal studies, but this creates serious problems for continuous monitoring of very small animals. Repeated blood drawings rapidly deplete the red blood cell supply. To avoid this, serum sampling can be done only infrequently. In addition, the procedure of collecting a serum sample is itself stressful and can interfere with subsequent behavior.

To alleviate these problems, we have made use of all the arboreal habits of our monkeys and of our naturalistic environments, as well as of the fact that most primates urinate upon waking early in the morning. We enter the environments as the animals wake up each morning and give food treats to those whose samples we want. We wait until they start to urinate and then hold a container underneath them to collect the sample without having to capture, handle, or restrain the animal. We have been able to follow females' reproductive physiology for 7–10 months, including sampling late in pregnancy, on the day of parturition, and during lactation, without compromising the health of the female or her infants. We can measure estrogens, luteinizing hormone, and chorionic gonadotropin (Ziegler,

Bridson, Snowdon, & Eman, 1987; Ziegler et al., 1987). With these techniques, we have been able to provide the most complete description of hormonal changes over a pregnancy cycle and during puberty of any primate species, including human beings. We have been able to modify these techniques to extract hormonal data from fecal samples, which allows us to study hormonal changes in wild populations.

Each of these noninvasive methods has allowed us to gather research data more efficiently or in more detail than would have been possible with traditional methods. The successful application of these noninvasive methods is a direct result of the more naturalistic environments that we have constructed.

SUMMARY

Our approach to psychological well-being has been influenced by working with endangered primates. Large, naturalistic habitats that mimic aspects of the natural habitat have led to high reproductive success, to animals that acquire the physical, cognitive, and social skills that would be useful if they were reintroduced to the wild, and to animals that can obtain a sense of control over the environment through moving through a complex environment and through interactions within a natural social group. The goals of much behavioral research on primates can best be met through large naturalistic environments, and the use of naturalistic environments has led to the development of new data-gathering techniques that are more efficient and can produce higher-quality data and yet are less invasive than many traditional techniques. Naturalistic environments can have a significant influence on psychological well-being.

References

Bauers, K. A., & Snowdon, C. T. (1990). Chirp discrimination in cotton-top tamarins. *American Journal of Primatology*, *21*, 53–60.

Brambell, M. R. (1977). Reintroduction. *International Zoo Yearbook*, *17*, 112–116.

Chamove, A. S., Anderson, J. R., Morgan-Jones, S. C., & Jones, S. P. (1982). Deep woodchip litter: Hygiene feeding and behavioral enhancement in eight primate species. *International Journal for the Study of Animal Problems*, *3*, 308–318.

Cleveland, J., & Snowdon, C. T. (1982). The complex vocal repertoire of the adult cotton-top tamarin (*Saguinus oedipus*). *Zeitschrift fuer Tierpsychologie*, *58*, 231–270.

Cleveland, J., & Snowdon, C. T. (1984). Social development during the first 20 weeks in the cotton-top tamarin, *Saguinus o. oedipus*. *Animal Behaviour*, *32*, 432–444.

Diener, E. (1984). Subjective well-being. *Psychological Bulletin*, *95*, 542–575.

Epple, G. (1978). Reproductive and social behavior of marmosets with special reference to captive breeding. *Primates in Medicine*, *10*, 50–62.

French, J. A., Abbott, D. H., & Snowdon, C. T. (1984). The effect of social environments on estrogen secretion, scent marking and sociosexual behavior in tamarins (*Saguinus oedipus*). *American Journal of Primatology*, *6*, 155–167.

French, J. A., & Inglett, B. J. (1989). Female-female aggression and male indifference in responses to unfamiliar intruders in lion tamarins. *Animal Behaviour*, *37*, 487–497.

French, J. A., & Snowdon, C. T. (1981). Sexual dimorphism in responses to unfamiliar intruders in tamarins (*Saguinus oedipus*). *Animal Behaviour*, *29*, 822–829.

French, J. A., & Stribley, J. A. (1987). Ovarian cycles are synchronized between and within social groups of lion tamarins (*Leontopithecus rosalia*). *American Journal of Primatology*, *12*, 469–478.

Gautier, J.-P., & Gautier-Hion, A. (1982). Vocal communication within a group of monkeys: Analysis by biotelemetry. In C. T. Snowdon, C. H. Brown, & M. R. Petersen (Eds.), *Primate communication* (pp. 5–29). New York: Cambridge University Press.

Gengozian, N., Batson, J. S., & Smith, T. A. (1978). Breeding of marmosets in a colony environment. *Primates in Medicine*, *10*, 71–78.

Hanson, J. P., Larson, M. E., & Snowdon, C. T. (1976). The effects of control over high intensity noise on plasma cortisol levels in rhesus monkeys. *Behavioral Biology*, *16*, 333–340.

Harlow, H. F. (1949). The formation of learning sets. *Psychological Review*, *56*, 51–56.

Hayes, S. L., & Snowdon, C. T. (1990). Predator recognition in cotton-top tamarins (*Saguinus oedipus*). *Animal Journal of Primatology*, *20*, 283–291.

Herzog, M., & Hopf, S. (1984). Behavioral responses to species-specific warning calls in infant squirrel monkeys reared in social isolation. *American Journal of Primatology*, *7*, 99–106.

Hoage, R. J. (1977). Parental care in *Leontopithecus rosalia*: Sex and age differences in carrying behavior and the role of prior experience. In D. G. Kleiman (Ed.), *The biology and conservation of the Callitrichidae* (pp. 293–305). Washington, DC: Smithsonian Institution Press.

Kleiman, D. G., Beck, B. B., Dietz, J. M., Dietz, L. A., Ballou, J. D., & Coimbra-Filho, A. F. (1986). Conservation program for the golden lion tamarin: Captive research and management, ecological studies, educational strategies and reintroduction. In K. Benirschke (Ed.), *Primates: The road to self-sustaining populations* (pp. 959–979). New York: Springer Verlag.

Konstant, W. R., & Mittermeier, R. A. (1982). Introduction, reintroduction and translocation of Neotropical primates: Past experiences and future possibilities. *International Zoo Yearbook*, *22*, 69–77.

McConnell, P. G., & Snowdon, C. T. (1986). Vocal interactions between unfamiliar groups of captive cotton-top tamarins. *Behaviour*, *97*, 273–296.

McGrew, W. C., Brennan, J. A., & Russell, J. (1986). An artificial "gum-tree" for marmosets (*Callithrix j. jacchus*). *Zoo Biology*, *5*, 45–50.

McGrew, W. C., & McLuckie, E. C. (1986). Philopatry and dispersion in the cotton-top tamarin, *Saguinus o. oedipus*: An attempted laboratory simulation. *International Journal of Primatology*, *7*, 401–422.

Miles, R. C., & Meyer, D. R. (1956). Learning sets in marmosets. *Journal of Comparative and Physiological Psychology*, *49*, 219–222.

Miller, D. B. (1977). Roles of naturalistic observation in comparative psychology. *American Psychologist*, *32*, 211–219.

Mineka, S., & Cook, M. (1988). Social learning and the acquisition of snake fear in monkeys. In T. Zentall, & B. G. Galef, Jr. (Eds.). Social learning: *Psychological and biological perspectives* (pp. 51–74). New Jersey: Lawrence Erlbaum Associates.

Mineka, S., Davidson, M., Cook, M., & Keir, R. (1984). Observational conditioning of snake fear in rhesus monkeys. *Journal of Abnormal Psychology*, *93*, 355–372.

Mineka, S., Gunnar, M., & Champoux, M. (1986). The effects of control in the early social and emotional development of rhesus monkeys. *Child Development*, *57*, 1241–1256.

Mineka, S., Keir, R., & Price, V. (1980). Fear of snakes in wild and lab-reared rhesus monkeys. *Animal Learning and Behavior*, *8*, 653–663.

Newman, J. D., & Symmes, D. (1982). Inheritance and experience in the acquisition of primate acoustic behavior. In C. T. Snowdon, C. H. Brown, & M. R. Peterson (Eds.). *Primate Communication* (pp. 259–278). New York: Cambridge University Press.

Neyman, P. F. (1977). Aspects of the ecology and social organization of free ranging cotton-top tamarins (*Saguinus oedipus*) and the conservation status of the species. In D. G. Kleiman (Ed.). *The biology and conservation of the Callitrichidae* (pp. 39–71). Washington, DC: Smithsonian Institution Press.

Novak, M. A., & Suomi, S. J. (1988). Psychological well-being of primates in captivity. *American Psychologist*, *43*, 765–773.

Petersen, M. R., Beecher, M. D., Zoloth, S. R., Green, S., Marler, P., Moody, D. B., & Stebbins, W. C. (1984). Neural lateralization of vocalizations by Japanese macaques: Communicative significance is more important than acoustic structure. *Behavioral Neuroscience*, 98, 779–790.

Plimpton, E. H., Swartz, K. B., & Rosenblum, L. A. (1981). The effects of foraging demand on social interactions in a laboratory group of bonnet macaques. *International Journal of Primatology*, *2*, 175–185.

Rosenblum, L. A., & Smiley, J. (1984). Therapeutic effects of an imposed foraging task in disturbed monkeys. *Journal of Child Psychology and Psychiatry*, *25*, 485–497.

Savage, A., Dronzek, L. A., & Snowdon, C. T. (1987). Color discrimination and its relation to the feeding ecology of the cotton-top tamarin (*Saguinus o. oedipus*). *Folia Primatologica*, *49*, 57–69.

Savage, A., Ziegler, T. E., & Snowdon, C. T. (1988). Sociosexual development, pair-bond formation, and mechanisms of fertility suppression in female cotton-top tamarins (*Saguinus oedipus oedipus*). *American Journal of Primatology*, *14*, 345–359.

Seyfarth, R. M., & Cheney, D. L. (1986). Vocal development in vervet monkeys. *Animal Behaviour*, *34*, 1640–1658.

Snowdon, C. T. (1989). The criteria for successful propagation of endangered primates. *Zoo Biology*, *Suppl. 1*, 149–161.

Snowdon, C. T., & Savage, A. (1989). Psychological well-being of captive primates: General considerations and examples from Callitrichidae. In E. Segal (Ed.). *Housing, care and psychological wellbeing of captive and laboratory primates* (pp. 75–88). Park Ridge, NJ: Noyes Publishing.

Snowdon, C. T., Savage, A., & McConnell, P. B. (1985). A breeding colony of cotton-top tamarins (*Saguinus oedipus*). *Laboratory Animal Science*, *35*, 477–481.

Tardif, S. D., Richter, C. B., & Carson, R. L. (1984a). Reproductive performance of three species of Callitrichidae. *Laboratory Animal Science*, *34*, 272–275.

Tardif, S. D., Richter, C. B., & Carson, R. L. (1984b). Effects of sibling rearing experience on future rearing success in two species of Callitrichidae. *American Journal of Primatology*, *6*, 377–380.

Ziegler, T. E., Bridson, W. E., Snowdon, C. T., & Eman, S. (1987). Urinary gonadotropin and estrogen excretion during the postpartum estrus, conception and pregnancy in the cotton-top tamarin (*Saguinus oedipus oedipus*). *American Journal of Primatology*, *12*, 127–140.

Ziegler, T. E., Savage, A., Scheffler, G., Snowdon, C. T. (1987). The endocrinology of puberty and reproductive functioning in female cotton-top tamarins (*Saguinus oedipus*) under varying social conditions. *Biology of Reproduction*, *37*, 618–627.

PART FOUR

IMPLICATIONS OF THE NEW AMENDMENTS FROM AN ADMINISTRATIVE AND A VETERINARY PERSPECTIVE

CHAPTER 13

PSYCHOLOGICAL WELL-BEING: THE BILLION-DOLLAR SOLUTION

THOMAS L. WOLFE

The 1985 amendment to the Animal Welfare Act uses the phrase "ensure the psychological well-being of nonhuman primates" to stipulate a requirement for assessment and provision of conditions necessary to provide nonhuman primates with a sense of well-being. In its wisdom, Congress left to the Department of Agriculture (USDA) the definition of what this term means in practice. Promulgation of the Animal Welfare Act (1985) has already taken the USDA more than 5 years, and implementation will impose an economic burden of perhaps a billion dollars on biomedical science and the U.S. taxpayer. Dozens of national and local seminars and workshops have sought to understand the term *psychological well-being* and how it might be implemented. Users and manufacturers of caging equipment and supplies feel in limbo, not certain what future regulations will require. At this point it seems that neither Congress nor the public appreciates what is at stake.

In the final analysis, what is the likelihood that the well-being of primates will be enhanced by this amendment? Is there a better way to achieve the goal of assuring the well-being of nonhuman primates than by means of vague legislation? History will say whether this legislation proves to be the most reasonable and cost-effective means of enhancing the well-being of animals, but in the meantime veterinarians, scientists, and institutions struggle with how to proceed. Using this legislation and existing guidelines and principles as examples, this paper will discuss these questions and offer suggestions for how to proceed.

NATIONAL COMMITMENT TO HUMANE CARE AND USE OF RESEARCH ANIMALS

The *Guide for the Care and Use of Laboratory Animals* (*1985*) has long stressed that humane care is a major responsibility of investigators and institutions using animals in research, testing, and education. In each of the five revisions since 1963, the *Guide* committees have been unwilling to attempt to write definitive care and use standards because of the great variability of research requirements and the differing physical and psychosocial needs of the many species of animals used in research; yet, sound, humane guidelines have evolved for use by knowledgeable professionals. Both the 1986 *U.S. Public Health Service Policy on Humane Care and Use of Laboratory Animals* (PHS Policy, 1986) and the 1985 *U.S. Government Principles for the Utilization and Care of Vertebrate Animals Used in Testing, Research and Training* (U.S. Government Principles, 1985) also place great responsibility for humane care and use of animals upon institutions and investigators, and likewise depend upon sound professional judgment. Based upon the intent and guidance of each of the documents, institutional programs are being established in response to societal goals of humane care and use. For the benefit of animals and of science, however, these programs should not be developed from a single blueprint. A fundamental requirement that institutions must impose is that programs and features introduced for the benefit of the animals be accomplished with an understanding of the needs of each animal and an acceptance of the costs and benefits of the intervention upon both the animals and the science. The congressional desire to provide for animals' well-being falls short of an understanding of animals' needs and risks excessive solutions that are not entirely in the best interest of the animals, the institutions, or biomedical science.

How then can institutions proceed? Under existing policies, institutions can initiate strategies to enhance their animals' well-being based on research and professional judgement and, when these strategies are found to be wrong, new hypotheses can be formed and new approaches tried. Much of laboratory animal science has progressed in just this manner. On the other hand, if such programs are based on laws that are found to be flawed, they are much more difficult to change and risk increasing the cost and slowing the pace of research while doing little for the well-being of animals, or worse, of actually doing them harm. Such laws, although motivated by good intentions, are likely to serve more to limit imaginative husbandry and enrichment strategies than to elucidate optimal methods. When few scientific principles exist, as in the case of understanding animal well-being, laws, regulations, and policies, when required, should clearly express intent but resist the temptation to create a blueprint for how the intent is to be achieved.

What is the role of the the Department of Agriculture? Considerable optimism exists regarding the leadership of the Animal and Plant Health Inspection Service (APHIS) today and its commitment to professionalism in animal welfare enforcement, with concomitant training of APHIS inspectors. APHIS now recognizes that inspectors charged with enforcement of animal welfare laws should be as well-trained and experienced as the investigators and veterinarians who are being called upon to implement the regulations. It is hoped that each animal welfare inspector, now freed from competing duties by a restructuring of APHIS, will be trained in the many nuances of laboratory animal welfare and will be able to evaluate an institution's animal well-being initiatives that are based upon professional judgment and research requirements as well as on the needs of the species and the individual animals. This capability requires considerable commitment from APHIS' entire animal welfare staff and a change in inspection policy from that of using a check-sheet of engineered standards to that of evaluating the performance of the specific institution.

Quite naturally, this policy will permit APHIS to make determinations about the welfare of animals based upon the animals' appearance and the performance of the institution in establishing and achieving reasonable goals of husbandry, care, and use. Because of the nature of the task, animal welfare will be better served by well-trained federal inspectors operating with professional judgment and fewer detailed regulations than by inspectors operating from rigidly designed standards on which the inspectors have, or exercise, little discretionary authority. Training, of course, is expensive, and the APHIS budget must be expanded to permit it. Indeed, if APHIS had originally been adequately funded to administer the Animal Welfare Act, it seems highly probable that Congress would not have sensed a necessity for new legislation.

FOCUSING RESPONSIBILITY

The lack of a clearly established, single focus of responsibility for the entire animal research program within each institution is perhaps the greatest impediment to developing and achieving optimal goals for care and housing. Well-being, enrichment, social housing, and exercise are important aspects of colony management, but their implementation depends upon sound institutional support and on an institutional program that will embrace them.

When a senior institutional official with authority to commit needed resources assumes responsibility for the animal care and use program and communicates this program to the faculty and the public, things can begin to happen. With responsibility clearly anchored, authority can be delegated to investigators, institutional veterinarians, and the Animal Care and Use Committees (ACUC) to develop and implement a formalized institutional program on animal care and use.

The institutional official, however, must remain central to the entire process and speak for the institution on these matters.

DEVELOPING THE INSTITUTIONAL PROGRAM

It is frequently the case that the veterinarian is given the responsibility, but not the authority, of ensuring that the institution maintains "assurance" with the Public Health Service and accreditation by the American Association for Accreditation of Laboratory Animal Care and that it avoids negative reports from the Department of Agriculture. With the realization that these are institutional responsibilities that cannot be delegated comes the understanding that they can only be accomplished when authority is duly delegated to those who can address broad institutional issues such as the space, humanpower, and financial resources that are required to achieve these goals.

Following identification of responsibility and delegation of authority, progress toward the goals of providing for the well-being of the animals and of developing an efficient, responsive, and accountable animal care and use program can begin. Each facet of the institutional program must be clearly established in a document that is signed by the institutional official and well understood by everyone involved. Such a program is normally developed by the ACUC with input and advice from veterinarians, investigators, medical and biosafety officers, technicians, engineers, budget and personnel officers, education specialists, and others. It should be considered as implementing a vital and changing policy, subject to inclusions or exclusions as new information becomes available.

Paramount in this policy is the development of a statement of the requirements for adequate veterinary care and authority for the institutional veterinarian to implement those requirements. Also, this program should define the roles of the institutional veterinarian, investigators, and the ACUC in review of research proposals. Each of the many facets of animal husbandry, colony management, and well-being should be covered. For example, issues identified in the *Guide* that should be considered include the creation of protocols on such topics as personnel training and health, animal procurement, environmental and social enrichment, husbandry, quality control, biohazards, disposition of carcasses and biological waste, pain control, and euthanasia.

In many institutions, such a program also establishes the relationship between the ACUC and the contractors and grantees that use or hold animals and seeks to establish long-range institutional goals, such as accreditation. Only through this type of comprehensive program can the true well-being of each animal be addressed. Without this institutional structure, a focus only on toys, space, enrichment, or exercise is short-sighted and not likely to succeed in the long run. The importance of this infrastructure far exceeds, yet often becomes subordinate to, the more easily identified issues of bricks and mortar and the compelling issues of "exercise" and "enrichment."

EDUCATIONAL PROGRAMS FOR INVESTIGATORS AND THEIR TECHNICIANS

The 1985 Animal Welfare Act amendment, *PHS Policy*, *Government Principles*, and the *Guide* all state that investigators and technicians should be trained to accomplish the goals of their protocols and to humanely handle the species to be used. Education is clearly the most expedient and long-lasting mechanism for addressing the issues of animal welfare. A popular construct for addressing these concerns is the 3-Rs of Russell and Birch: *reduction* of pain and distress: *replacement* of animals with nonanimal alternatives when consistent with the scientific goals; and *refinement* of experimental design to minimize the number of animals required. Investigators should be instructed in the application of each "R" to his or her research and to the planning of future research. These educational programs should also discuss common research techniques for each species, such as surgery, restraint, venipuncture, anesthesia, analgesia, and euthanasia. The exact content of each program will vary greatly between institutions, from providing videotapes for viewing by investigators to mandatory lectures and hands-on training.

Whatever the format, the goals of the education and the requirements placed upon investigators and technicians should be established in the overall institutional program and communicated by the senior institutional official. As in the institutional program itself, the educational and training program must be allowed to change and evolve over time. Two committees of the Institute of Laboratory Animal Resources, National Research Council are currently addressing in detail the topics of education of investigators and technicians and of pain and distress in animals. These reports will be published by the National Academy of Sciences.

ESTABLISHING PSYCHOLOGICAL WELL-BEING AND ENVIRONMENTAL ENRICHMENT

The choice of the term *psychological well-being* is criticized as being too vague to be meaningful and as serving only to waste time, energy, and money as individuals and institutions seek to understand how to define and implement it for each species. Perhaps this is true. Yet, because of the dearth of knowledge about the subject, it may prove beneficial that this ambiguous law was cast, rather than one for which all the parameters were known and specified. The requirement, to provide a physical environment to promote the psychological well-being of research animals, is an opportunity to review current practices and implement specific goals to enhance the quality of laboratory animal care. Research institutions have sensed the spirit of Congress and seized the opportunity. Responding with liberal quantities of professional judgment and with few ironclad rules, they have made significant strides toward their goals. In the absence of narrowly promulgated regulations, even more progress is possible.

It is to be expected that some institutions will evolve strategies for enrichment that differ from those recommended in the *Guide* or required by the Animal Welfare Act regulations. Federal and private inspectors and their parent agencies or organizations must respect this process and know when to resist "going by the book." This kind of inspection, of course, will require that federal inspectors and accreditation site visitors be adequately trained and sensitized to know when a policy or regulation is on thin ice and when the institutional initiatives—especially those that differ from recommended or mandated practices—merit a trial. This inspector-institution relationship must be allowed considerable freedom and must not be constrained by excessive waiver procedures. Negotiation between institutions and inspectors for trials of innovative techniques should become the norm, not the exception. Quite diverse but equally beneficial approaches to the problem of enrichment will be implemented by different institutions, and the best practices will evolve in time from these multiple approaches. Approached in this manner, it may be that as the Animal Welfare Act amendment is transformed into regulations, its real value will be judged to be due to its impetus in stimulating a sense of common commitment between research facilities and regulating agencies and to its heuristic statement of intent.

ASSESSING THE WELL-BEING OF ANIMALS

The principal problem with attempting to develop programs to achieve *psychological well-being* is that the two words in this term are each unnecessarily limiting, undefinable in the context, and provide no insight into how to measure or achieve the state desired. *Psychological* imparts only confusion and misdirection and adds nothing to *well-being* used alone. Psychologists, philosophers, and physiologists since before Descartes and Pavlov have been trying to understand the psyche of animals. We are not likely to add significantly to this literature. We would do well to stop speaking about the animal's psyche when what we mean is its well-being.

Because of the difficulty of measuring well-being, its assessment is often through its antithesis, distress—used here to mean stress over which the animal has neither innate or learned control and for which it lacks appropriate coping and avoidance responses. This definition focuses on the stimulus, which may or may not be harmful but which is always *perceived* as aversive. (Even a kind, gentle technician is perceived as a threat to a dog, if that dog is unsocialized. The dog is then in a state of distress.) The definition suggests that not all stress is bad. In fact, most animals, including humans, seek some form of stress to relieve an otherwise sedentary life. Clearly, minimization or avoidance of all stress is not the goal.

Distress should be differentiated from "eustress" or innocuous stress, with which animals effectively cope and might even seek and enjoy. Distress can be thought of as stress that an animal would generally, but not invariably, choose to avoid or escape and that is accompanied by nonproductive or abnormal behavioral or physiological manifestations.

For normal development of man or beast, stress plays an important role. Without carefully designed stresses, especially during early development, normally innocuous events later in life (like conspecific or human social contacts, or restraint) may be perceived by laboratory animals as highly distressing. Early stressful experiences are important for the well-being of the animals and for the validity of the science.

Quite simply, whether the experience is to be a human contact, restraint, a new social group, or any novel experience, well-being protocols should provide for adaptation of the animal to the new situation. Adaptation teaches coping strategies that eliminate distress. Also, at least for some animals, early socialization to people and to conspecifics serves as an important bridge (reducing distress) to the ability to cope with exposure to quite unrelated novel situations later in life.

Common sense and many research articles tell us that dogs reared without human contact are fearful and distressed when in close proximity to people. When this fear results in aggressive or abnormal behavior, or in alterations in body physiology or chemistry, the dog is clearly distressed, and the research to which it is assigned can be profoundly affected. However appealing it may seem to humans, to an animal not accustomed to it, human contact is distressing indeed! And as for other stressors, the animal must be taught appropriate coping strategies.

Without adaptation and a knowledge of the early experiences of the animals, interpretation of any data derived during the course of, for example, close human interaction, might be misleading. This source of possible error adds an important variable in our understanding of well-being. Because of the likelihood that animals in a study have different previous experiences, it becomes obvious that what is pleasing for one can be highly distressful for another. In this case, a single species-specific approach to well-being would be effective for only part of the population.

Well-being is not easily measurable: A simple analysis of cortisol, or any other single measure, can often be deceptive and cannot be considered the definitive marker of distress. Cortisol does not discriminate among subjective (psychological) states and may become elevated under a variety of situations, including intense pleasure as well as pain or distress.

In order to get around this problem, scientists commonly use such indicators of well-being as food and water consumption, weight gain or loss, general appearance, reproduction, and parenting. While these are critical parameters to assess, they too lack specificity in that such measures may simply serve to provide a ranking of the importance of the measure to the animal.

For example, assessment of a pair of mice in a standard "shoe box"-type cage may indicate that food consumption and reproduction are normal. The conclusion that these animals are in a desired state of well-being may be premature, however, for the observations may simply reflect that hunger and reproductive drives are sufficiently strong to override the shortcomings of the caging and thus say little about its perceived quality. In spite of these deficiencies, however, behavior and clinical signs of health are likely to be the best and most important keys to assessment of well-being by well trained, observant individuals.

ASSESSING WELL-BEING

Well-being is often thought to be provided through environmental enrichment, both physical and social. But, as discussed, assessment of the value of such enrichment to the animal is most difficult. Perhaps, enrichment should be considered as anything that the animal finds pleasing at that moment in time. Enrichment might include provision of human interaction, a cagemate, a nesting box, opportunity to forage for food in deep bedding or work for food through manipulation of "food puzzles," or control over some aspect of the environment, such as visual access to other animals, temperature, music, or video. Of course, selection of the presumed enrichment must be made intelligently and with a rather detailed knowledge of the species' normal behavior and the individual's previous experience.

Placing a solitary species in a social group, giving a tree limb to a ground-dweller or to a burrowing animal, or making any sudden change in habitat to which the animal has not become accustomed cannot be expected to produce the desired state of well-being. In this regard, close human contact with an unsocialized animal, or conversely, placing a highly human-oriented animal in exercise pens removed from human activity, would not be likely to be ranked very desirable by the animals themselves.

WHEN IN DOUBT, LISTEN TO THE ANIMALS

It is not difficult to devise enrichment strategies, but it is exceedingly difficult to know with certainty how the animal perceives them. One exceedingly simple technique is through preference tests in which the animal is given access to two cages with slightly different furnishings or opportunities. The question posed to the animal is, "Which cage configuration do you prefer?" If the animal answers by spending 90% of the time in one cage, interpretation that that is the "preferred cage" is probably valid—but only for that configuration and at that time.

One cannot be immediately certain, however, whether this selection indicates evidence of aversiveness to the alternative (e.g., eye contact by an adjacent dominant animal) or a true preference. Perhaps the animal in this situation is saying, "Given these choices, I prefer cage A." But if the options are changed, the social structure in the room is changed, or an animal with a different previous experience is tested, the preferences may change.

Even to a novice, the responses of a social colony of cats to different rooms with identical space would be obvious. Consider two rooms, one including vertical space with perches at different heights, the other providing only horizontal space and all perches at the same level. The preferences of cats to seek heights would immediately suggest something of possible importance.

Through a battery of choices and a knowledge of species normative behavior, preference tests may provide an understanding of true preferences. Their use seems like a reasonable and responsible approach that can be taken by any institution with little or no additional resources. Whether enrichment is offered through preference tests or controlled introduction of toys, manipulanda, or food puzzles, the assessment of enrichment should start with a thorough knowledge of the animal's behavior and early experience. Evaluations should include a combination of selected biochemical, physiological, and behavioral measures, and a good understanding of the strengths and limitations of each.

This volume concentrates on the well-being of nonhuman primates, yet this paper has introduced examples using dogs, cats, and rodents. This was done for comparative purposes and to pose a question. For many years, macaques were imported from the wild, and their aggression was considered a desirable indicator of normal behavior. Yet, we know that when these animals are adequately adapted to people and the environment, they and their baseline physiological values change considerably.

Today increasingly greater percentages of macaques and other primates are born in domestic breeding colonies, providing greater opportunities to select for disease- and carrier-free states (e.g., simian herpes and simian immunodeficiency viruses). With them is also the opportunity to initiate behavioral training and socialization in young animals.

Zoonoses make direct human contact with some species inadvisable today, but it is not unlikely that within a few years this obstacle to behavior training will be overcome, and more and more primates will enter laboratories previously trained to drink from medicine cups, present arms or legs for venipuncture, urinate, or enter a transfer cage on command. Dog or cat breeders today would not consider raising offspring unsocialized to people. With elimination of the fear of zoonoses, and improved screening and diagnostic tests, will there remain reason to not do the same for nonhuman primates in the future?

WHERE TO FROM HERE?

This discussion has served to address the strengths and weaknesses of expressing, through legislation, a societal concern for the care of research animals. Because of the complexities of studying, understanding, recognizing, and providing well-being in the many animal species used in research, it seems that narrowly promulgated regulations requiring the provision of psychological well-being will serve little constructive purpose. However, it is likely that history will look favorably on the process stimulated by the law. Every indication suggests that regulated institutions and implementing inspectors will have the opportunity to discuss approaches that are in the best interest of the animals and that bricks, mortar, and puddles will assume a more realistic place in the priorities of animal welfare.

What is clear is that the issues of humane care and well-being are genuine and that each institution that uses animals should establish a timetable, develop an institutional plan, and address the issues seriously. The perceived need for future legislation may well depend upon how we respond today.

References

Animal Welfare Act, 7 U.S.C. §§ 2131-2157 (1985).

National Research Council. (1985). *Guide for the care and use of laboratory animals.* (NIH Pub. No. 85-23). Washington, DC: U.S. Department of Health and Human Services.

Public Health Service (1986). *Public Health Service on humane care and use of laboratory animals.* Washington, DC: U.S. Department of Health and Human Services.

U.S. Government Principles. (1985). Principles for the utilization and care of vertebrate animals used in testing, research, and training. In *Guide for the care and use of laboratory animals* pp 82–83. (NIH Pub. No. 86-23). Washington, DC: U.S. Department of Health and Human Services.

CHAPTER 14

A PRESCRIPTION FOR PSYCHOLOGICAL WELL-BEING

ROSALIND M. ROLLAND

Much discussion, little consensus of opinion, and no definite answers characterize the continuing debate over the meaning of the phrase *psychological well-being* as introduced by Congress in the proposed Animal Welfare Act (1985) and enacted as Public Law 99–148. Primate specialists in a variety of disciplines do agree that currently there is no precise definition of "psychological well-being" as it pertains to captive nonhuman primates (Novak & Suomi, 1988; Petto, Novak, Fingold, & Walsh, 1989; Woolverton, Ator, Beardsley, & Carroll, 1989). Furthermore, objective quantifiable measurements that accurately reflect the psychological well-being of a nonhuman primate have yet to be developed and accepted (see C.L. Coe, chapter 10; L. Johnson, A. Petto, & P.K. Sehgal, chapter 11; S. Line, H. Markowitz, K.N. Morgan, & S. Strong, chapter 17; G.P. Sackett, chapter 5; and S.J. Suomi & M.A. Novak, chapter 7; in this volume). The problem posed by these new regulations will be how to promote psychological well-being when one can neither define it nor measure it objectively.

Assuming that regulations proposed by the Animal and Plant Health Inspection Service of the United States Department of Agriculture (USDA) become implemented, research institutions housing nonhuman primates will have to face the practical considerations of compliance. The requirements for group housing, exercise periods, social interaction, and environmental enrichment will have a major impact upon colony management, disease control, and the design of research projects. Although the objective of increasing the well-being and humane treatment of captive primates is undeniably indicative of a good trend, institution of these changes may not always benefit the health of a given primate. The costs and benefits must be weighed, and a decision must be made on behalf of each individual (e.g., M. Keeling, P. Alford, & M. Bloomsmith, chapter 8; and S.J. Suomi & M.A. Novak, chapter 7; in this volume).

THE ROLE OF THE CLINICAL VETERINARIAN

Who, then, will balance the positives and negatives and arrive at a prescription for the psychological well-being of each captive primate? As the proposed regulations are now written, it appears that the "attending veterinarian" of the research institution will bear this responsibility. Section 3.81 of Part III of the amendments delineates the specific requirements that must be instituted to promote psychological well-being (USDA, 1990). Individual exemptions from these requirements can only be granted by the attending veterinarian and can be invoked only "because of their health or condition, or in consideration of their well-being." Exemptions may also be granted by the institutional Animal Care and Use Committee "for scientific reasons set forth in the research proposal." Furthermore, records of social interaction, exercise periods, environmental enrichment devices, and exemptions must be kept by the research facility and are subject to inspection by USDA officials or any other federal funding agency. Therefore, it seems inevitable that compliance with these regulations and decisions regarding the psychological state of individual primates will fall primarily upon the clinical veterinarian.

How are veterinarians who work with nonhuman primates prepared to meet this new challenge? Veterinary clinical training focuses predominantly upon the diagnosis, treatment, and prevention of disease. Most veterinary work concerns physical health: prevention of disease with vaccinations, diagnosis of disease with laboratory tests, and treatment of disease with pharmacologics. Although mental health is undeniably an integral component of well-being in general, the recognition and treatment of problems is far less objective (see R.D. Hunt, chapter 3 in this volume). There is no single blood test to measure the state of the psyche; there is no vaccine to prevent psychological ailments; and there is no pharmacologic agent with which to treat victims of psychological distress. Few clinical veterinarians have formal training in primate behavior and observational techniques. Promoting psychological well-being will involve a different set of skills from those generally acquired during even specialized veterinary training. Although the intent to promote psychological well-being may be genuine, the background training and resources necessary for success may not be readily available to most clinical veterinarians.

Aside from a lack of formal training in primate behavior, veterinarians will also face a new demand upon personnel resources. Determination of the social compatibility of a group of nonhuman primates requires many hours of observation to define the social dynamics. Detailed records of group interactions and individual social histories will have to be maintained. Fulfilling exercise requirements for singly caged primates will require personnel to transfer animals to exercise cages, keep records, and sanitize these areas between uses. For a facility housing a large number of primates, this represents an enormous investment of personnel hours.

Most veterinarians will be facing a considerable challenge to implement successfully these proposed regulations. Implementation will require at least rudimentary knowledge of primate behavior and observational techniques, both for veterinarians and for animal-care personnel. It will necessitate development of strategies to determine social compatibility and protocols for the transition from single caging to group housing (e.g, M. Keeling et al., chapter 8 in this volume). Some objective criteria of well-being, both physical and mental, will need to be determined to facilitate rational decisions on behalf of each individual. An efficient record system will be necessary to document social histories, exemptions from group housing, exercise periods, and social interaction for each primate. Without a significant commitment to provide the resources needed to formulate a psychological well-being program at each research institution, implementation of these changes will be haphazard, at best.

IMPLICATIONS FOR PHYSICAL HEALTH

The requirements for social grouping and exercise will doubtless benefit the physical health of many individual animals. Problems associated with singly caged primates, such as obesity, muscle atrophy, and self-mutilation should decline in frequency. Although impossible to quantify, the quality of life of captive nonhuman primates will probably improve in group housing. For all the potential benefits of these changes, certain individual animals are likely to suffer physically from the effort to promote psychological well-being.

Group housing can be physically traumatic and sometimes life-threatening. By far the most common physical problem that I treat as a clinical veterinarian is trauma sustained by macaques in group-housing situations. This occurs even when no changes have been made within a previously compatible group. The incidence of traumatic injuries is increased when new groups are formed or when animals have to be removed and reintroduced into a group for medical reasons, as inevitably occurs. Trauma may range from superficial abrasions to multiple wounds and lacerations, sometimes leading to life-threatening loss of blood and shock.

The risk of physical injury may be an acceptable one for most animals. However, even in the absence of wounds, lower ranking macaques in any group experience a chronic physical stress of a different nature. Competition for food and shelter is a given in a social-housing situation, and compatibility may change as social alliances are formed (Chapais, 1988). Lower ranking individuals bear the brunt of the competition and of changes in social dynamics. The result can be malnutrition and suboptimal bodyweights for these individuals. During cold weather in northern climates, if low-ranking primates are excluded from heated shelters, serious frostbite or hypothermia can result (Petto, 1988). Chronic social stress probably has other physical ramifications, such as an effect on immune system function (see C.L. Coe, chapter 10 in this volume), yet to be identified. In some cases, a single cage may be both physically and psychologically preferable to group housing.

Restrictions on social grouping are listed in section 3.86 of Part III of the proposed regulations (USDA, 1990). Part III states that "non-human primates may not be housed with other species of primates or animals unless they are compatible, do not prevent access to food, water, or shelter by individual animals, and are not known to be hazardous to the health and well-being of each other." (p. 33525). Theoretically, all singly caged primates could be exempted from the social housing requirement on the basis of this statement. For example, take the fact that competition for food and shelter and changes in compatibility are to be expected in all social groups. Therefore, there is always the possibility of fighting; socially housed primates are always a threat to one another. Although the benefits of group housing are great, a unique set of costs in the form of stresses and physical risk is inherent in group-housing situations.

Social housing and interaction will also affect the efficiency of disease prevention, diagnosis, and treatment. The recognition and treatment of physical problems is facilitated by single caging and obscured by group housing. Subtle changes associated with ill health such as anorexia, lethargy, or a change in water consumption may be readily apparent in the singly housed primate, but very difficult to discern in a group situation. In general, the signs of disease must be more dramatic and severe to be recognizable in individuals housed in a group. Frequently, recognition is associated with a more advanced stage of the disease process. For example, puncture wounds resulting from bites are often difficult to identify in primates housed in social groups. The injury may not be recognized until the wound has become infected, resulting in abscess formation or a discharge that can be seen. The delay in wound care and antibiotic therapy will certainly have an impact on the health and well-being of these individuals.

In group housing, outbreaks of infectious diseases such as *Shigellosis* and *Campylobacter* infection are less easily controlled and will involve larger numbers of primates as the organism is passed between members of the group. Even with rigorous sanitation procedures, the use of a common exercise area for primates that remain singly housed increases the possibility of transmission of infectious diseases. It would be impossible to sterilize these areas totally between uses. Therefore, from a strictly medical viewpoint, the singly caged primate is likely to receive better care, as diseases and problems can be readily recognized and treated in the early stages. Group housing will result in a certain compromise in veterinary care to the individual.

Although many individuals will benefit from the transition to group housing, others may not fare so well. For example, in the 8 months from April through November, 1989, there were 57 injuries requiring removal from group housing and treatment in the clinic in our group-housed, timed mating colony of about 120 females. Thirty-seven animals were treated and released into the social group without further incident, but eight of the animals (18% of the total number injured) account for 35% of the total injuries recorded. Even though most had had previous social experience, it is clear that at least some of the animals were less well off in these groups of breeding females than they had been in individual cages.

Implementation of the requirements for group housing and social interaction will necessitate weighing potential benefits against the potential costs to each individual. At least initially, there will undoubtedly be an increase in physical trauma cases as groups are formed from previously singly caged animals. We have seen such trauma in our new program for group housing of timed-mated rhesus monkeys. Early diagnosis and treatment of disease will be less efficient, which also could increase the morbidity and probably the mortality rate. Infectious disease outbreaks may be more difficult to control.

THE PROGNOSIS FOR PSYCHOLOGICAL WELL-BEING

Even with these potential difficulties, the movement towards providing a more stimulating environment for captive nonhuman primates is a very worthwhile endeavor. Everyone involved in primate research will benefit from healthier research subjects. Adjustments will have to be made in colony management and research protocols taking into account the new housing regulations. An increase in the number of trauma cases, and probably an increase in morbidity and mortality from various disease processes, may have to be accepted. Nevertheless, the goal to improve the existence of captive primates will be worth these costs if it is realized for the majority.

However, will compliance with the specific requirements for providing for psychological well-being necessarily accomplish this goal? This system will be difficult to evaluate until some objective criteria for defining the psychological condition of nonhuman primates are developed. However, most important will be the commitment that each research institution is willing to make towards designing its own program for promotion of psychological well-being. If the onus of implementation is merely dumped on the clinical veterinarian with no additional resources to develop such a program, the potential benefits will be greatly diminished. As there is no universal prescription for psychological well-being, new approaches to colony management, housing, facility design, and environmental enrichment will have to be tried. Without the support necessary to test and evaluate alternatives to current management practices, the success of any implemented changes in promoting the psychological well-being of captive primates may be purely serendipitous.

References

Chapais, B. (1988). Rank maintenance in female japanese macaques: Experimental evidence for social dependency. *Behavior, 104*, 41–59.

Novak, M. A., & Suomi, S. J. (1988). Psychological well-being of primates in captivity. *American Psychologist, 43*, 765–773.

Petto, A. J. (1988). "Fall Fever" in Lemur Catta: Near death from an agonistic encounter between adult males. *Canadian Review of Physical Anthropology, 6*, 31–34.

Petto, A. J., Novak, M. A., Fingold, S. A., & Walsh, A. C. (1989). The search for psychological well-being in nonhuman primates: Information resources. *Science and Technology Libraries, 10*, 101–127.

U.S. Department of Agriculture, Animal and Plant Health Inspection Service, Animal Welfare; Standards; Proposed Rule (9 C.F.R., Part 3) [Docket No. 90-040]. Federal Register 55 (158), 33448-33531; August 15, 1990.

Woolverton, W. L., Ator, M. A., Beardsley, P. M., & Carroll, M. E. (1989). Effects of environmental conditions on the psychological well-being of primates: A review of the literature. *Life Sciences, 44*, 901–917.

PART FIVE

PROCEDURES FOR PROMOTING PSYCHOLOGICAL WELL-BEING IN NONHUMAN PRIMATES

CHAPTER 15

EFFORTS TO PROMOTE PSYCHOLOGICAL WELL-BEING IN PROSIMIAN PRIMATES AT THE DUKE UNIVERSITY PRIMATE RESEARCH CENTER

M. KAY IZARD

The Duke University Primate Center (DUPC) is a research institution and a captive conservation and breeding center for prosimian primates (Izard, 1989). The DUPC houses over 650 prosimian primates representing six families and 33 species and subspecies (Table 1). Over 70% of the colony consists of endangered lemurs from Madagascar. The taxonomic diversity of the collection is reflected in a wide variety of locomotor adaptations, body weights, diets, reproductive patterns, and social organizations. As one example, among the prosimians housed at the DUPC, gestation length ranges from 60 days in the tiny mouse lemur (*Microcebus murinus*) to 192 days in the slow loris (*Nycticebus coucang*) (Figure 1).

To meet the research objectives of the DUPC, our goal is to house healthy animals with normal patterns of species-specific behavior. To meet our conservation and breeding objectives, our goal is to provide husbandry conditions in which animals with many different life-history strategies will reproduce successfully. To assess whether our conservation and breeding objectives are being met,

I thank Phyllis Arrington for manuscript preparation and Steve Daugherty for the photographs. This is paper no. 454 of the DUPC. This work was partially supported by NSF BSR 85-21235.

Table 1

PROSIMIAN SPECIES HOUSED AT THE DUKE UNIVERSITY PRIMATE RESEARCH CENTER

LEMURIFORMES
Lemur fulvus fulvus (Brown lemur)
Lemur fulvus albifrons (White-fronted lemur)
Lemur fulvus collaris (Collared lemur)
Lemur fulvus rufus (Red-fronted lemur)
Lemur fulvus sanfordi (Sanford's lemur)
Lemur macaco macaco (Black lemur)
Lemur macaco flavifrons (Sclater's lemur)
Lemur catta (Ringtailed lemur)
Lemur rubiventer (Red-bellied lemur)
Lemur mongoz (Mongoose lemur)
Lemur coronatus (Crowned lemur)
Varecia variegata variegata (Black & White ruffed lemur)
Varecia variegata rubra (Red ruffed lemur)
Hapalemur griseus (Grey gentle lemur)
Propithecus verreauxi (Sifaka)
Propithecus tattersalli (Golden crowned sifaka)
Cheirogaleus major (Greater dwarf lemur)
Cheirogaleus medius (Fat-tailed dwarf lemur)
Mirza coquereli (Greater mouse lemur)
Microcebus murinus (Lesser mouse lemur)
Daubentonia madagascariensis (Aye-aye)
LORISIFORMES
Perodicticus potto (Potto)
Nycticebus coucang (Slow loris)
Nycticebus pygmaeus (Pygmy slow loris)
Loris tardigradus (Slender loris)
Galago demidovii (Demidoff's bushbaby)
Galago senegalensis moholi (Lesser bushbaby)
Galago garnettii (Brown bushbaby)
Galago crassicaudatus monteiri (Thick-tailed bushbaby)
Galago crassicaudatus argentatus (Thick-tailed bushbaby)
Galago crassicaudatus crassicaudatus (Thick-tailed bushbaby)
TARSIIFORMES
Tarsius syrichta (Phillipine tarsier)
Tarsius bancanus (Bornean tarsier)

we ask the following questions: Are the females cycling? Do the males have spermatozoa? Are the animals copulating? Are the females conceiving? Are they carrying their infants to term? Are they lactating? Are they behaving maternally? Are they raising their infants to maturity? If the answer is no to any of the above questions, then we vary three facets of captive husbandry (housing, diet, social groupings) in an attempt to promote successful captive reproduction.

In our attempts to promote successful reproduction in captivity, we address psychological well-being as a combination of both behavioral and physiological

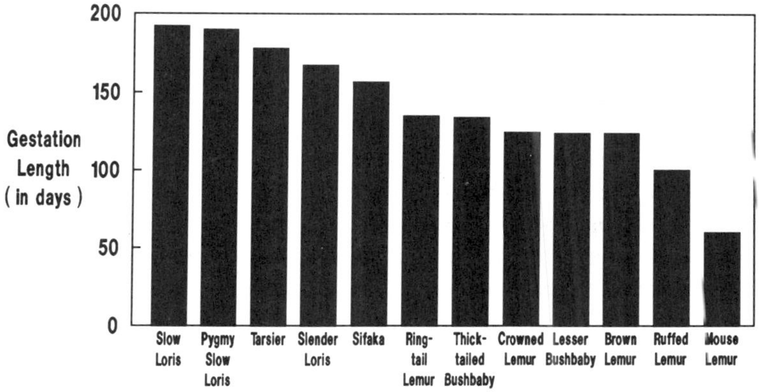

Figure 1 Gestation lengths of some prosimian primates, illustrating the diversity of the DUPC group.

variables. Our approach is similar to that delineated by Sackett (see chapter 5 in this volume), that is, we choose a set of functions and find a structure to produce them. We keep trying until we find the structure, as reflected in housing, diet, and social grouping, that yields successful reproduction in captivity.

HOUSING

There are four major types of housing at the DUPC. These include indoor cages, indoor rooms, outdoor cages, and outdoor roofless enclosures.

Indoor Cages

All of the nocturnal prosimians at the DUPC are housed in indoor cages or rooms. With the exception of the aye-aye (*Daubentonia madagascariensis*), the nocturnal prosimians all weigh less than 2.5 kg. Because of their small size, they would be vulnerable to predation if kept outdoors. In addition, their nocturnal activity pattern makes outdoor housing inconvenient to the human observers of their behavior. Indoor housing allows partial reversal of light cycles so that the nocturnal activity period can occur during the normal waking hours of humans. Partial reversal of the light cycle allows feeding and cleaning to be done in the morning when the animals are inactive and observation of the animals for the purposes of research and husbandry to be done in the afternoon and evening.

In the indoor cages, we provide furnishings that enhance the complexity of the environment while meeting the diverse needs of our nocturnal species. For example, nestboxes are provided for species that keep their infants in nests, like the lesser bushbaby, *Galago senegalensis* (Bearder & Doyle, 1974), the mouse lemurs (Martin, 1972), and the dwarf lemur (*Cheirogaleus medius*) (Foerg, 1982), and suitable perches are included for species that "park" their infants, like the slow loris (Ehrlich, 1974).

Lesser bushbabies, which weigh 180–200 g, are maintained in cages of sufficient vertical height to enable them to display their vertical clinging and leaping mode of locomotion. Nestboxes are provided for sleeping as well as for use as nests for infants. Shelves and ladders of washable vinyl-coated wire add complexity to the cage. The cages are constructed so that small compartments can be closed off from the main cage to isolate sick or pregnant animals.

The greater bushbaby (*Galago crassicaudatus*), which weighs about 2 kg, is housed in cages divided into three vertically arranged compartments that triple the available floor space for the animals. Shelves and nestboxes add additional complexity to the cage. The mouse lemur and dwarf lemur are provided with vine loops in their cage interiors. These form continuous runways for these animals, which are poor jumpers. These lemurs are also provided with nestbottles and nestboxes for sleeping and for nesting infants.

Indoor Rooms

The indoor rooms at the DUPC are more naturalistic than are the indoor cages, because animals in these rooms are frequently used for behavioral observations. The naturalistic cage furnishings promote the full species-typical range of behaviors, including scent-marking behavior, which is of paramount importance in the olfactory communication system of prosimian primates (Schilling, 1979).

The dimensions of the substrates in each room are tailored to the body size and locomotor specializations of each species. Mouse lemurs are housed in indoor rooms furnished with vine tangles of small dimensions to accommodate their small body size. The greater mouse lemur (*Mirza coquereli*), which weighs five times as much as the mouse lemur, is provided with larger trees and branches. The slender loris (*Loris tardigradus*), slow loris (*Nycticebus coucang*), and potto (*Perodicticus potto*) cannot jump and therefore are provided with continuous pathways to food, water, and resting shelves. These species must also have a branch situated so that they can assume the upside down copulatory posture that is typical of the lorisines (Izard & Rasmussen, 1985; Izard, Weisenseel, & Ange, 1988). Failure to provide such an opportunity has probably resulted in the historically poor reproductive success of these species in captivity.

Two species of tarsier (*Tarsius bancanus* and *T. syrichta*) are housed at the DUPC. Rooms for tarsiers, which are small vertical clingers and leapers, can easily be furnished with bamboo of varying diameters to provide vertical uprights for locomotion. Other bamboo pieces are attached to the uprights at 45–90° angles to stabilize the interior substrates, to provide forks for resting, and for locomotion of the prey items of the tarsier.

The materials that we use at the DUPC to furnish our indoor rooms and cages are primarily natural materials. These are renewable resources, like kudzu, bamboo, grapevines, and honeysuckle. These materials can be removed, discarded, and replaced on a regular basis.

Although a variety of plastic nestboxes is used in indoor rooms and cages, many of our nestboxes are constructed of wood. The wood used in indoor cages and rooms is treated with polyurethane or painted to make its surface impervious to moisture. The advantages of wood furnishings are that this material stimulates scent-marking and olfactory communication, provides a secure grip, and is aesthetically pleasing, not only to the caretakers of the animals, but to potential donors to our conservation programs.

Outdoor Cages

A number of indoor rooms at the DUPC have outdoor access. These are used for sifakas (*Propithecus verreauxi*) and bamboo lemurs (*Hapalemur griseus*). These animals are relatively large vertical clingers and leapers and therefore are provided with large vertical supports which, in most cases, are cedar trees that have been treated with polyurethane. Macedonia (1987) examined the effects of housing variables on activity budgets in sifakas. Animals housed outdoors were significantly more active and spent more time in locomotion, feeding, and playing than animals housed indoors. When given a choice between indoor and outdoor housing, sifakas chose to be outdoors even if their outdoor run was smaller than their indoor room. One sifaka group with access to an outdoor run spent only 8% of its time in its larger indoor compartment, even though fed in both compartments. Seventy-five percent of the time spent indoors was spent eating or sitting under heat lamps on cool mornings. These observations support other findings (Erwin, 1979; Novak & Drewsen, 1989) that more space is not necessarily better. Quality of space may well be more important than quantity.

With the exception of the sifaka and the bamboo lemurs, all of the other diurnal lemurs at the DUPC are kept outdoors year round. Many of the outdoor cages are covered by overhangs or other structures, but many are roofed only with cage wire. The cages are all provided with shelter from rain and sun. Because North Carolina occasionally has snow, ice, and very cold weather, insulated nestboxes are provided, fitted with heatlamps controlled by a rheostat (Figure 2). If the outdoor temperature falls below 10°C (50°F), the heatlamps come on automatically. Multiple nestboxes are provided for each social group to prevent one

Figure 2 An outdoor cage at the DUPC, with heated nestbox and cedar tree.

Figure 3 An outdoor cage at the DUPC, depicting substrates of varying orientations.

animal from monopolizing the source of heat. One side is made of Plexiglas to allow caretakers to monitor animals visually when they are inside the nestbox.

The outdoor cages contain trees, vines, and branches, which provide substrates of varying orientations for locomotion (Figure 3). They also provide opportunities for spur marking, an activity performed frequently by the ring-tailed lemur (*Lemur catta*) and the bamboo lemur. Some of the outdoor cages were erected around stands of bamboo. There is some wear and tear on the bamboo from the activity of the animals in the cage, but if the cage is left empty for several weeks, the bamboo grows back quickly.

By providing naturalistic housing, we increase opportunities for the performance of natural behaviors by the animals. Our naturalistic housing makes the DUPC attractive to those interested in studying the behavior of captive primates.

Figure 4 A natural-habitat enclosure at the DUPC, with Electronet electrified fencing on top of chain-link fencing and a cleared swath on either side of fence.

By providing housing that is not barren and uniform, we promote both behavioral research and captive propagation.

Outdoor Roofless Enclosures

Perhaps the most attractive features of the DUPC for research and conservation are the outdoor roofless enclosures or natural habitat enclosures. Four of these enclosures, encompassing 27 acres, house seven species and subspecies of diurnal lemurs. Each enclosure contains mixed hardwood and conifer forest representing a wide variety of North Carolina trees and plants. The enclosures are bounded by chain-link fence topped with Electronet (Premier Fence Systems, Figure 4). A

5m-wide swath of trees is cleared on each side of the fence to prevent the animals from escaping.

Arboreal and terrestrial quadrupeds, like the ring-tailed lemur and the ruffed lemur (*Varecia variegata*), have opportunities to exercise their considerable climbing and jumping skills in the natural habitat enclosures. Smaller cages within the enclosures allow animals to be confined close to heated nestboxes in freezing weather. These confinement cages are also used to catch the animals on a regular basis. Thus, research that requires frequent handling of animals while maintaining them in a social setting is possible at the DUPC.

The forest provides an infinite variety of locomotor options and opportunities for the expression of close to the full range of natural behaviors. In these enclosures, we have observed sifakas doing the bipedal sideways hop, which had previously been observed only in the wild. In these enclosures ruffed lemurs, which are the only diurnal lemurs that do not carry their infants clinging to their bodies, have built ground nests of branches in which to have their infants (Pereira, Klepper, & Simons, 1987). Nestbuilding has never been observed in our caged ruffed lemurs.

We are fortunate that prosimians are gentle on their environments. We can keep them in cages that a monkey would tear apart. The outdoor enclosures show little effect of lemur habitation, even after almost 8 years of occupancy in one enclosure. The kinds of caging improvements we make are inexpensive and do not significantly raise the cost of research.

Unfortunately, what does raise the cost of research is trying to comply with sanitation requirements. The heavy emphasis that the Animal Welfare Act places on surfaces that can be sanitized threatens to curtail our efforts to provide housing that meets the needs of the animals by promoting captive conservation and reproduction. The new regulations proposed by the Animal and Plant Health Inspection Service, however, recognize the importance of naturalistic housing for behavioral research.

DIET

Diet at the DUPC is tailored to each species' preferences and adaptations. The diet for each species is varied daily. A variety of fruits and vegetables is delivered weekly, as are crickets, lizards, and grubs. Purina Monkey Chow is also delivered weekly and forms the staple diet of all species except the tarsiers and aye-ayes. A mealworm colony is maintained at the DUPC so that mealworms are available daily.

Tarsiers are carnivorous and eat only live animal prey. They must be provided with live crickets, lizards, and mealworms that are accessible. Because the tarsiers will not come to the ground to feed, live prey are released twice daily onto the bamboo and branches located within the rooms. Cages made of wire are inappropriate for tarsiers because the live prey can escape through the cage sides. The rooms are misted twice daily to provide humidity for the tarsiers and drinking water for the crickets and lizards.

For frugivorous species, like the crowned lemur (*Lemur coronatus*) and the six subspecies of *Lemur fulvus*, a different fruit is provided each day along with the daily ration of Monkey Chow. These fruits vary seasonally according to availability, but include apples, bananas, oranges, cantaloupes, watermelon, grapes, pears, and peaches.

The highly folivorous sifakas are fed mango leaves daily, as well as a variety of native North Carolina foliage, including maple, hornbeam, redbud, sweetgum, and sumac. By housing the sifakas outdoors, adding North Carolina foliage to their diets and freezing browse for winter use, we have begun to see the birth of healthy full-term babies that are surviving to adulthood (Haring, 1988). Bamboo lemurs are also folivorous, but need a high percentage of their diet supplied in the form of a certain type of bamboo. Since we have been allowing the bamboo lemurs access to the outdoors, we have also seen the birth of healthy infants to this species—one that has always been difficult to breed in captivity.

The mongoose lemur (*Lemur mongoz*) is another prosimian that historically has been difficult to breed in captivity. Up until several years ago, there was very little captive propagation of this species, even though the captive population was fairly large. Wild-caught animals would produce one or two infants and then stop breeding (Schaaf & Stuart, 1983). There were virtually no captive-born mongoose lemurs breeding at all.

All of the animals we had at the DUPC prior to 1982 were acquired from zoos. Without exception these animals were obese. In 1982, we acquired several pairs of wild-caught mongoose lemurs from the Comore Islands. On the diet we normally feed to lemurs of the same body weight, these wild-caught mongoose lemurs began gaining weight and were verging on obesity. To solve the obesity problem, we put our mongoose lemurs on a diet that included low-calorie vegetables and smaller quantities of Monkey Chow. In the last 4 years with the new diet, we have had 18 infants born. Both wild-caught and captive-born animals have bred. The extent to which successful reproduction can be traced directly to diet, to other variables alone, or in combination is now being investigated.

SOCIAL GROUPS

Social organization in prosimian primates varies from that of the solitary aye-aye to the complex multigenerational matriarchies of the ring-tailed lemur (Bearder, 1986; Richard, 1986). The nocturnal prosimians tend to have simple types of social organizations, usually with the territory of one male overlapping those of several females and their offspring. To approximate this social system, most nocturnal prosimians at the DUPC are kept in pairs with their most recent offspring. At times, though, it may be advantageous to disrupt this arrangement.

In three species of *Galago* at the DUPC, the neonatal mortality rate is about 60% if females give birth while caged with another animal. If the female is isolated, neonatal mortality drops significantly to around 20% (Izard & Simons, 1986). Reproductive success can be improved by manipulating the social group at the time of parturition. This change in social group may resemble what wild pregnant *Galago* initiate themselves. Although galagos normally sleep during the day with a small group in a nest or tree cavity, pregnant females leave the group prior to parturition and do not come back until several weeks later (Sauer, 1967).

The diurnal lemur species are typically found in larger social groups. We house them in family groups that may include the offspring of a mated pair from the last 3 or 4 years. Under these conditions, usually only the original female will reproduce. In ruffed lemurs (*Varecia variegata*), males housed with their fathers are physiologically inhibited, and their testicles remain small during the breeding season when the testicles of uninhibited males enlarge (Foerg & Hoffman, 1982). Groups are carefully monitored at the onset of the breeding season to detect instances of rejection before they escalate into situations where severe wounding will occur. The daughters are set up in new social groups after being rejected from their natal group.

The social groups of all of our species are monitored carefully for signs of aggression, such as chases, bite wounds, and peripheralization of group members, all of which precede expulsion from the social group (Vick & Pereira, 1989). These behaviors are considered to be normal behaviors, but we must respond with the drastic measure of removing the targeted animal from the social group to prevent severe injury to the rejected animals.

CONCLUSION

At the DUPC, we exert efforts to obtain our research and conservation goals by providing for the animals' physical well-being and social needs through manipulation of housing, diet, and social groups. We can measure behavior and reproductive output to see if our efforts are having a positive effect. Although we cannot measure feelings of happiness, contentment, or psychological well-being in animals that do not speak our language, we believe that if the physical and social needs of the animals are met so that they behave normally and reproduce successfully in captivity, their psychological well-being is enhanced and promoted. Our techniques enable us to make major contributions to the captive conservation and study of prosimian primates at a time when captive conservation of primates is imperative.

References

Bearder, S. K. (1986). Lorises, bushbabies, and tarsiers: Diverse societies in solitary foragers. In A. B. Smuts, D. L. Cheney, R. M. Seyfarth, R. W. Wrangham, & T. T. Struhsaker (Eds.), *Primate societies* (pp. 11–24). Chicago: The University of Chicago Press.

Bearder, S. K., & Doyle, G. A. (1974). Ecology of bushbabies, *G. senegalensis* and *G. crassicaudatus*, with some notes on their behavior in the field. In R. D. Martin, G. A. Doyle, & A. C. Walker (Eds.), *Prosimian biology* (pp. 109–130). London: Duckworth and Company, Ltd.

Ehrlich, A. (1974). Infant development in two prosimian species: Greater galago and slow loris. *Developmental Psychobiology, 7*, 439–454.

Erwin, J. (1979). Aggression in captive macaques: Interaction of social and spatial factors. In J. Erwin, T. Maple, & G. Mitchell (Eds.), *Captivity and behavior: Primates in breeding colonies, laboratories, and zoos* (pp. 139–171). New York: Van Nostrand Reinhold.

Foerg, R. (1982). Reproductive behavior in *Varecia variegata. Folia Primatologica, 38*, 108–121.

Foerg, R., & Hoffman, R. (1982). Reproduction in *Cheirogaleus medius. Folia Primatologica, 39*, 49–62.

Haring, D. (1988). Natural history and captive management of Verreaux's sifaka. *International Zoo Yearbook, 27*, 125–134.

Izard, M. K. (1989). The Duke University Primate Center. *Journal of Medical Primatology, 18*, 85–98.

Izard, M. K., & Rasmussen, D. T. (1985). Reproduction in the slender loris (*Loris tardigradus malibaricus*). *American Journal of Primatology, 8*, 153–165.

Izard, M. K., & Simons, E. L. (1986). Isolation of females prior to parturition reduces neonatal mortality in *Galago. American Journal of Primatology, 10*, 249–255.

Izard, M. K., Weisenseel, K. A., & Ange, R. L. (1988). Reproduction in the slow loris (*Nycticebus coucang*). *American Journal of Primatology, 16*, 331–339.

Macedonia, J. (1987). Effects of housing differences upon activity budgets in captive sifakas (*Propithecus verreauxi*). *Zoo Biology, 6*, 55–67.

Martin, R. D. (1972). A preliminary field study of the lesser mouse lemur (*Microcebus murinus*). (J. F. Miller 1777). *Zietschrift fur Tierpsychologie*, Suppl. 9, 42–90.

Novak, M. A., & Drewson, K. H. (1989). Enriching the lives of captive primates: Issues and problems. In E. F. Segal (Ed.), *Housing, care and psychological well-being of captive and laboratory primates* (pp. 161–182). New Jersey: Noyes Publications.

Pereira, M. E., Klepper, A., & Simons, E. L. (1987). Tactics of care for young infants by forest-living ruffed lemurs (*Varecia variegata variegata*): Ground nests, parking, and biparental guarding. *American Journal of Primatology, 13*, 129–144.

Richard, A. F. (1986). Malagasy prosimians: Female dominance. In B. B. Smuts, D. L. Cheney, R. M. Seyfarth, R. W. Wrangham, & T. T. Struhsaker (Eds.), *Primate societies* (pp. 25–33). Chicago: The University of Chicago Press.

Sauer, E. G. F. (1967). Mother-infant relationship in galagos and the oral child-transport among primates. *Folia Primatologica, 7*, 127–149.

Schaaf, C. D., & Stuart, M. D. (1983). Reproduction of the mongoose lemur (*Lemur mongoz*) in captivity. *Zoo Biology, 2*, 23–28.

Schilling, A. (1979). Olfactory communication in prosimians. In G. A. Doyle & R. D. Martin (Eds.), *The study of prosimian behavior* (pp. 461–538). New York: Academic Press.

Vick, L. G., & Pereira, M. E. (1989). Episodic targeting aggression and the history of *Lemur* social groups. *Behavioral Ecology and Sociobiology, 25*, 3–12.

CHAPTER 16

AN ENVIRONMENTAL ENRICHMENT PROGRAM FOR CAGED RHESUS MONKEYS AT THE WISCONSIN REGIONAL PRIMATE RESEARCH CENTER

VIKTOR REINHARDT

It is a contradiction to demand an objective definition for a subjective phenomenon such as the experience of psychological well-being of another being. However, acknowledging this principle, indisputable similarities between ourselves and nonhuman primate subjects (e.g., we are both intelligent mammals adapted to a complex inanimate and social environment) justify drawing basic inferences from our own experience in a given situation to the subject's experience in the same situation. For example, being continually confined in a room or cage with plenty of food but with no access to manipulable objects or to social partners is a distressing experience for human primates and, by inference, for nonhuman pri-

I am very thankful to Douglas Cowley, Steve Eisele, and Russel Vertein for their excellent assistance. Without their expertise and goodwill it would have been impossible to test and implement the various cage-enrichment strategies. I am also grateful to Robert Goy and Dan Houser for their encouraging interest in the enrichment program. Thanks are also due to Gregory Armstrong for granting me permission to obtain tree branches from the arboretum of the University of Wisconsin without charge. Special thanks are due to my wife, Annie Reinhardt, for editing this manuscript and to Robert Dodsworth for preparing the photographs of Figures 6 and 8 The environmental enrichment program is supported by NIH grant RR00167 to the Wisconsin Regional Primate Research Center.

Figure 1 Rhesus monkeys confined in barren cages, without opportunities to express their social disposition, are understimulated.

mates as well (Figure 1). Anyone with an informed understanding of the nature of primates in general, and of nonhuman primates specifically, will not even raise the question of whether the psychological well-being of nonhuman primates living alone in barren cages is enhanced by providing them with species-adequate stimulation (environmental enrichment).

In 1986, these considerations prompted systematic efforts at the Wisconsin Regional Primate Research Center (WRPRC) to explore ways of enriching the environment of individually caged rhesus monkeys (*Macaca mulatta*). Several options proved to be practicable and were implemented as a management routine.

Figure 2 A segment of a deciduous tree branch is better than nothing to provide some stimulation.

INANIMATE OBJECTS

Loose segments of deciduous tree branches were used as objects for gnawing and manipulating (Figure 2). Deciduous tree branches and polyvinyl chloride pipes were used as perches (Figure 3). The lower end of a branch or pipe is attached with a chain at the front of the cage while its upper end rests at the junction of the back and side walls without touching the top grating. The perch is suspended high enough so that an animal can both sit on it and freely move below it; usable cage space is thus increased.

Figure 3 Deciduous tree branches provide inexpensive perches and, if properly installed, also increase the available cage space for the animals.

During a census at the WRPRC in August 1989, there were 619 individually caged rhesus monkeys older than 1 year. Without exception, all of them were provided with a loose branch segment, with a perch, or with both. After 1 year of exposure, 92 animals of both sexes and all ages were observed for 1 hour. Individuals were seen on this occasion using their branch segments (for gnawing, manipulating, rolling, or carrying around) in 67% of cases and using their perches (for sitting, holding, shaking, or swinging) in 91% of cases. The animals spent an average of 3% of the time with the branch segments and 31% with the perches (cf. Reinhardt, 1989a). Since 1986, more than 600 animals have been exposed to one or several of the these inanimate objects. None of them has suffered any injury or disease specifically related to an enrichment object.

Branches and branch segments are obtained free of charge. The cost per polyvinyl chloride pipe is about $2 and the attachment equipment per perch about $1.50. Perches and branch segments are cleaned with warm water daily and disinfected once every 2 weeks. It is not necessary to remove them from the cages during these procedures. Branches and branch segments wear out and need to be replaced regularly.

FACILITATED SOCIALIZATION

Adult-Infant Pairs

Individually caged adult rhesus monkeys (older than 4 years) of both sexes are paired in their home cages with naturally weaned surplus infants (12–18 months old) from the WRPRC's breeding colony. Male adult-infant pairs are moved to male-only areas (no visual contact with females) when the young are 3 years old to avoid the risk of injury involved in potential sexual competition. Of 91 pairs tested, partners were compatible (adequate food sharing, no serious injuries, no signs of depression) in 88% (80 out of 91) of the cases throughout follow-up periods of 4 months to 14 years (Figure 4). Compatibility was not dependent on the partner's sex (Reinhardt, Houser, Eisele, & Champoux, 1987).

The companions of 9 pairs were observed for one hour each, 2 1/2 years after pair formation. All partners interacted in species-typical ways on an average of 23% of the time (14 out of 60 min.) (cf. Reinhardt, 1990a) (Figure 5).

Adult Same-Sex Pairs

Individually caged adults of both sexes are paired with same-sexed partners in double cages after dominance-subordination relationships have been established during a period of noncontact familiarization in a different double cage with grated partition. Male pairs are kept in strictly male-only areas to avoid antagonism resulting from sexual competition. Of 85 pairs tested, partners were compatible in 84% of the cases (71 out of 85) throughout follow-up periods of 3 months to 14½ years. The compatibility of female pairs did not differ from that of male pairs (Reinhardt, Houser, Eisele, Cowley, & Vertein, 1988; Reinhardt, 1989b).

Eleven female pairs were observed for one hour each, 2½ years after pair formation. All partners interacted with each other in species-typical ways approximately 31% of the time (19 out of 60 min.). Four male pairs also interacted with each other in species-typical ways on an average of 15% of the time (9 out of 60 min.) (cf. Reinhardt, 1990a).

Figure 4 Twelve-year old George protectively holds his infant companion, Billy, shortly after pair formation. Billy shows no signs of depression.

Figure 5 Twelve-year old Bruce and his 4-year old companion, Bill, were paired 2½ years ago. They have not lost interest in each other during that time, but regularly groom each other and play with each other.

Other Pairs

Individually caged, adult castrated males are paired in double cages that are new to both partners. Individually caged adult females are paired with males, and individually caged young animals (1–4 years old) are paired with each other in the same way. Of 5 adult castrated male pairs tested, partner compatibility was ascertained in 100% of cases throughout a follow-up period of 3 years. Of 12 adult female-male pairs and 30 pairs of young animals, partners were compatible in 92% and 93% of cases, respectively, throughout a follow-up period of 1½ years.

Altogether, 223 pairs were tested. Partners were compatible in 195 cases (87%); they were incompatible and had to be re-paired with other animals in 28 cases (13%). Incompatibility was related to depression and/or inadequate food sharing in 25 cases (11%); it was related to serious but not life-threatening injuries in 3 cases (1%).

Beginning in 1986, more than 400 animals were exposed to compatible companions. No indication was found that this increased the animals' susceptibility

Figure 6 Facilitated socialization does not interfere with tethering for remote infusion.

to infections, distress, or any other disease (Reinhardt, 1990b, 1990c). Paired animals do not require extra space, but each partner profits from the double-sized cage. Paired caging has no adverse impact on common research protocols such as tethering (Figure 6), venipuncture in home cage (Figure 7), timed breeding, and headcap implantation (Figure 8). If an animal has to be restrained in a chair during an experiment, its companion is kept nearby for psychological support (Reinhardt, Hauser, Cowley, Eisele, & Verkin, 1989).

At the August 1989 census of the WRPRC's caged rhesus monkeys, 62% (382 out of 619) lived in pairs. There were 40 adult female-infant pairs, 7 adult male-infant pairs, 75 adult female pairs, 18 adult male pairs, 4 adult castrated male pairs, 33 adult female-adult male pairs, and 14 pairs of young animals.

The environmental enrichment program described in this chapter has proven to be both effective over extended time and practical with no adverse impact on the routine management of the animals. This program shows that "where there's a will, there's a way" to enrich the environment of caged rhesus monkeys without undue risks to the animals or costs to the facility.

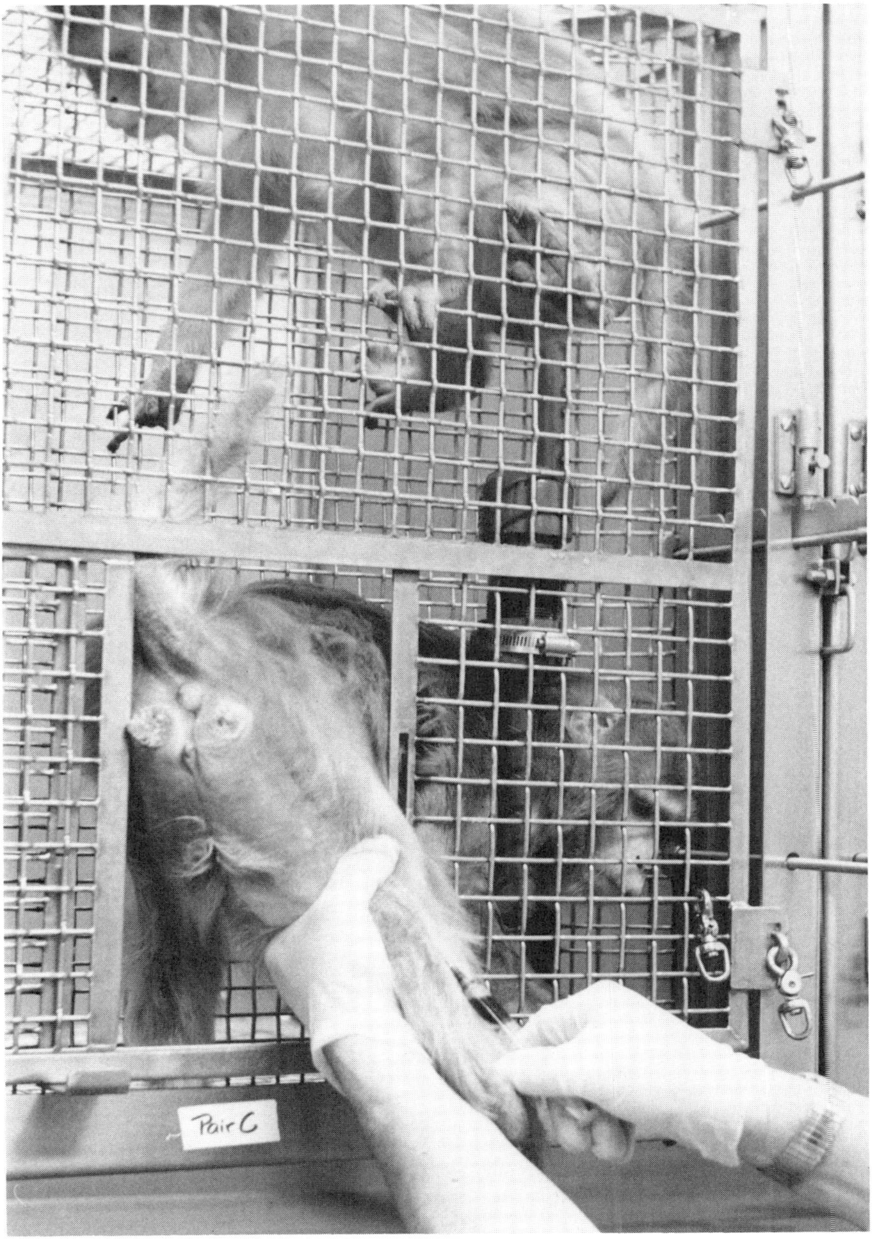

Figure 7 In their home cage, paired rhesus monkeys are readily trained to present a leg for venipuncture (Vertein & Reinhardt, 1989; Reinhardt, in press).

Figure 8 Pair-housing allows adherence to research protocols requiring headcap implantation.

References

Reinhardt, V., Houser, W., Eisele, S., & Champoux, M. (1987). Social enrichment with infants of the environment for singly caged adult rhesus monkeys. *Zoo Biology*, *6*, 365–371.

Reinhardt, V., Houser, W., Eisele, S., Cowley, D., & Vertein, R. (1988). Behavior responses of unrelated rhesus monkey females paired for the purpose of environmental enrichment. *American Journal of Primatology*, *14*, 135–140.

Reinhardt, V. (1989a). Evaluation of the long-term effectiveness of two environmental enrichment objects for singly caged rhesus macaques. *Lab Animal*, *18*, 31–33.

Reinhardt, V. (1989b). Behavioral responses of unrelated adult male rhesus monkeys familiarized and paired for the purpose of environmental enrichment. *American Journal of Primatology*, *17*, 243–248.

Reinhardt, V. (1990a). Time budget of caged rhesus monkeys exposed to a companion, a PVC perch and a piece of wood for an extended time. *American Journal of Primatology*, *20*, 51–56.
Reinhardt, V. (1990b). Social enrichment for laboratory primates: A critical review. *Laboratory Primate Newsletter*, *29*, 7–11.
Reinhardt, V. (1990c). Living continuously with a compatible companion is not a distressing experience for rhesus monkeys. *Laboratory Primate Newsletter*, *29*, 16–17.
Reinhardt, V. (1991). Training adult male rhesus monkeys to actively cooperate during in-homecage venipuncture. *Animal Technology*, *42*, in press.
Reinhardt, V., Houser, D., Cowley, D., Eisele, S., & Vertein, R. (1989). Alternatives to single caging of rhesus monkeys (Macaca mulatta) used in research. *Zeitschrift fuer Versuchstierkunde*, *32*, 275–279.
Vertein, R., & Reinhardt, V. (1989). Training female rhesus monkeys to cooperate during in-homecage venipuncture. *Laboratory Primate Newsletter*, *28*, 1–3.

CHAPTER 17

EFFECTS OF CAGE SIZE AND ENVIRONMENTAL ENRICHMENT ON BEHAVIORAL AND PHYSIOLOGICAL RESPONSES OF RHESUS MACAQUES TO THE STRESS OF DAILY EVENTS

SCOTT W. LINE, HAL MARKOWITZ, KATHLEEN N. MORGAN, AND SHARON STRONG

PSYCHOLOGICAL WELL-BEING

Laboratory animal scientists have long been concerned with improving animal care. In the past, research has focused primarily on evaluation of physical aspects of the laboratory environment, resulting in many improvements in health, nutrition, and safety. A logical extension of this line of inquiry is to explore the aspects of

This research was supported by NIH Grant RR00169 to the California Regional Primate Research Center. We thank Dave Hess, Oregon Regional Primate Research Center, for performing the cortisol assays, Mark Nakazono, Bill Wolden, and Grant Buckman for assistance in behavioral data collection, and Warren Miller of Willoughby Electronics for designing and building the door monitor panel.

the environment that influence the behavior of laboratory animals, with the goal of improving the nature of environmental impacts on the animals. Increasing public interest in this topic resulted in the 1985 amendment to the federal Animal Welfare Act mandating the provision of nonhuman primates with environments that promote their "psychological well-being."

The attempt to define psychological well-being has been difficult. The fact that it has taken over 5 years to develop a set of proposed regulations reflects this difficulty, which still exists even though attention to the effects of the environment on well-being is not a new phenomenon.

Scientists who use nonhuman animals in research have long known that impoverished environments have adverse effects on the physical health and behavioral suitability of their subjects. By the end of the 1950s, Heine Hediger had published several descriptions of behavioral and physiological pathologies found in captive animals. His suggestions for eliminating these problems included enclosures with more animal-usable space, social companions or suitable surrogates for naturally social species, and "furniture" that allows expression of a species' natural behavioral repertoire (Hediger, 1964). Attempting to evaluate objectively the effectiveness of these modifications in producing the desired improvements in animal welfare has not been easy. The effects of applying Hediger's suggestions to captive environments have been recorded for a variety of species, but a consensual definition of psychological well-being has still not appeared (Line, 1987; Novak & Suomi, 1988; Markowitz & Line, 1989).

Although the difficulty in obtaining agreement on a general definition of primate psychological well-being must be acknowledged, the use of objective measures, such as patterns of normal and abnormal behavior and patterns of physiological responses to challenges common in the laboratory environment, seems an obvious place to start. What we have attempted to do in our research with rhesus monkeys is establish operational definitions of well-being, based on changes in behavioral and autonomic nervous system responses. We believe that decreases in aberrant or self-abusive behavior, changes in amounts of normal behavior in the direction of those displayed by free-ranging animals and decreases in the magnitude of physiological responses to stressors are all evidence of increased well-being in laboratory-housed primates. We have used these definitions to see which, if any, of the environmental changes we make have measurable effects. Although we do not mean to suggest that the status quo is adequate, we do assert that changes in procedure should be based on sound, empirical evidence that they will have the desired effect, and not merely on intuitive suppositions.

We realize that our current view of what constitutes increased well-being for these animals will not be appropriate for all primates, all macaques, or even all rhesus monkeys. There is enormous diversity within the order Primates, and a definition of psychological well-being broad enough to apply to all species will necessarily be very general and perhaps of questionable utility. Animals of different ages, sexes, developmental backgrounds, and species will all need different things to ensure their well-being. Still, we cannot allow the obstacles to forming a general definition to prevent us from doing anything to improve laboratory conditions.

The definitions we are using are starting points for discussion, ones with the advantage of being based on objective observation and analysis and on some understanding of animals' responses to current laboratory life. By using these measurements, we hope to provide data of benefit to many institutions that need to make cost-effective decisions regarding the new requirements for promoting the psychological well-being of primates. Knowing something specific about the relative effects of factors such as cage size and enrichment devices in alleviating the effects of stressors in a laboratory will allow researchers and administrators to make more informed decisions concerning the ways that they can stretch limited budgets and improve the behavioral conditions of their primates. This chapter examines several aspects of the psychological well-being of laboratory primates: how singly caged rhesus macaques (*Macaca mulatta*) respond to (a) routine husbandry procedures, (b) changes in cage size, and (c) the opportunity to control some aspects of their environment through the use of an enrichment device.

For the past 3 years, we have worked to establish an active program in the assessment of psychological well-being at the California Regional Primate Research Center. We have concentrated on singly caged macaques, since single-caging is a common housing arrangement and the one that is most different from the environment in which these animals evolved. During that time, we have evaluated the effects of a number of passive and active devices placed in or on cages for use by rhesus monkeys (Clarke, Line, Ellman, & Markowitz, 1987; Line, Clarke, Ellman, & Markowitz, 1987; Line, Clarke, Markowitz, & Ellman, 1990; Line, Markowitz, Morgan, & Strong, 1989; Markowitz and Line, 1989).

The effects of environmental manipulations have been assessed by recording a combination of behavioral and physiological variables including plasma cortisol and heart rate. We have also examined the effects of varying cage size on these measures (Line, Morgan, Markowitz, & Strong, 1989a) and developed a system to routinely assess some of the changes that occur when various types of human work are accomplished in the animal's home quarters (Line, Morgan, Markowitz, & Strong, 1989b).

ROUTINE HUSBANDRY EVENTS

The first aspects of psychological well-being to be discussed are the physiological and behavioral responses of the primates to routine events such as feeding, cleaning, and cage changing that are performed by the animal care staff. Human activities are the major source of variation in the laboratory environment, and the monkeys are very attentive to people in the room. People are directly involved in events both pleasurable and aversive to the animals and thus have a significant impact on their well-being.

The following results were obtained during 6 months of observation of six adult female rhesus monkeys. They were individually housed in stainless steel squeeze-back cages in a room with approximately 40 other rhesus monkeys. Visual and auditory contact occurred with other monkeys across a central aisle, though physical contact with neighbors was extremely limited by the solid sides of the cages. The animals were fed a commercial diet twice a day; water was continually available. Lighting was automatically controlled on a 12:12 light:dark schedule.

A telemetry system[1] recorded the heart rate and amount of movement of all subjects at 2-minute intervals and allowed us to record remotely any acute changes in response to external events. Each monkey had a transmitter surgically implanted subcutaneously along the dorsal midline. The transmitter broadcast a radio signal encoding the electrocardiogram to a receiver mounted on the top of the subject's cage. Variations in the signal strength occurred as the monkey moved and were translated into a relative index of activity by the receiver. A computer in an adjacent room recorded the signals from each receiver. Telemetry data were collected from 1000 to 1700 hours 4 days per week and for 24 hours continuously 1 day per week. These data were subsequently averaged at 1-hour intervals for each day.

Human activity in the room was recorded by a panel mounted on the door. The panel contained 10 buttons listing different types of personnel and activities (feeding, cleaning, health check, or other animal technician activity; research technician; veterinarian; animal health technician; behavioral observation, blood sample, or other project staff activity). People were asked to press the appropriate button as they entered the room. A magnetic switch registered when the door opened and closed. Data from the door panel and the magnetic switch were also recorded at 2-minute intervals by the telemetry computer. These data allowed us to assess heart rate and activity of the monkeys in conjunction with different types of human activity in the room.

There were significant variations in both heart rate and movement during the 12 hours the lights were on (repeated measures analysis of variance [ANOVA], $F = 19.52$, $p < .0001$, Figure 1). Heart rate increased rapidly after the lights came on at 0600 to a peak of 150 beats per minute (bpm) between 0700 and 0800. This was the hour when the morning feeding took place. Activity also increased when the lights came on and peaked between 0800 and 0900. The daily cleaning took place during this hour. Both heart rate and activity decreased until 1100, then rose to a second peak between 1300 and 1400. The afternoon feeding took place around the time of the second peak. Activity gradually declined from this

[1]Manufactured by Mini-mitter Co., Inc., Sunriver, OR.

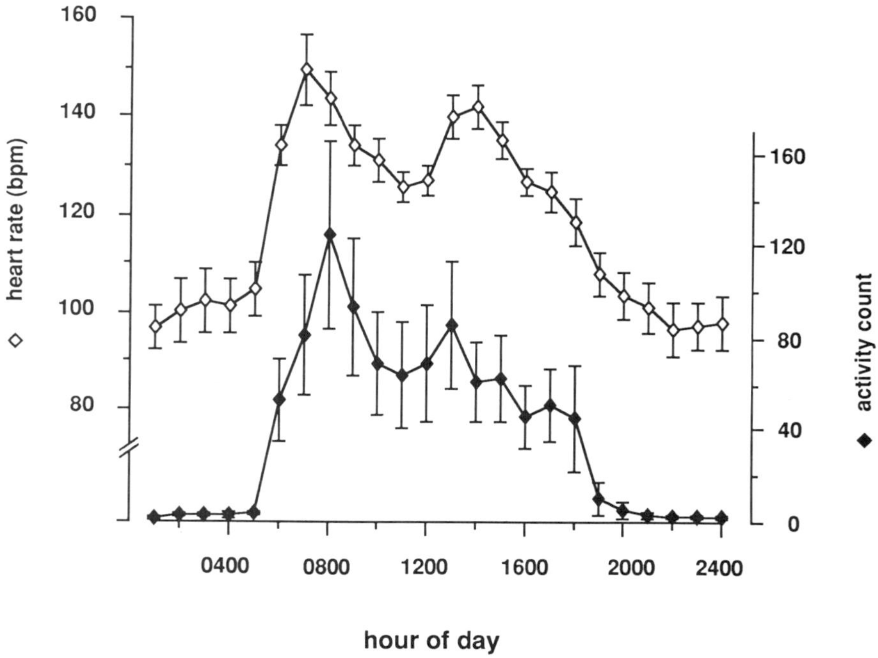

Figure 1 Mean daily cycles of heart rate and activity of six singly caged adult female rhesus monkeys. Each point represents the mean (± SEM) of 30 samples taken during a 1-hour period for 14 different days. Room lights were on from 0600 to 1800.

point until the lights went off at 1800 and then rapidly dropped to very low levels for the rest of the night. Heart rate gradually declined from 140 bpm at the second feeding to a low of 95 bpm at 2200 hours. It remained low until the lights came on the next day. These diurnal patterns were very stable and were observed throughout the data-collection period.

In order to assess how tightly linked these changes in heart rate and activity were to the human activities in the room, as opposed to time of day alone, we examined these variables immediately prior to and after human activity. Changes in activity level and heart rate occurred as soon as a technician entered to feed, clean, perform routine health checks, or administer treatments at night to other monkeys in the room (Figure 2). Activity increased during the health check, cleaning, and night treatment, and decreased during the afternoon feeding. Heart rate did not change during the health check, increased slightly during morning cleaning, declined during the afternoon feeding, and increased briefly during the night treatment. Despite the fact that these activities occurred daily, the monkeys continued to react to them. These responses were generally brief, however, and disappeared shortly after the person left the room.

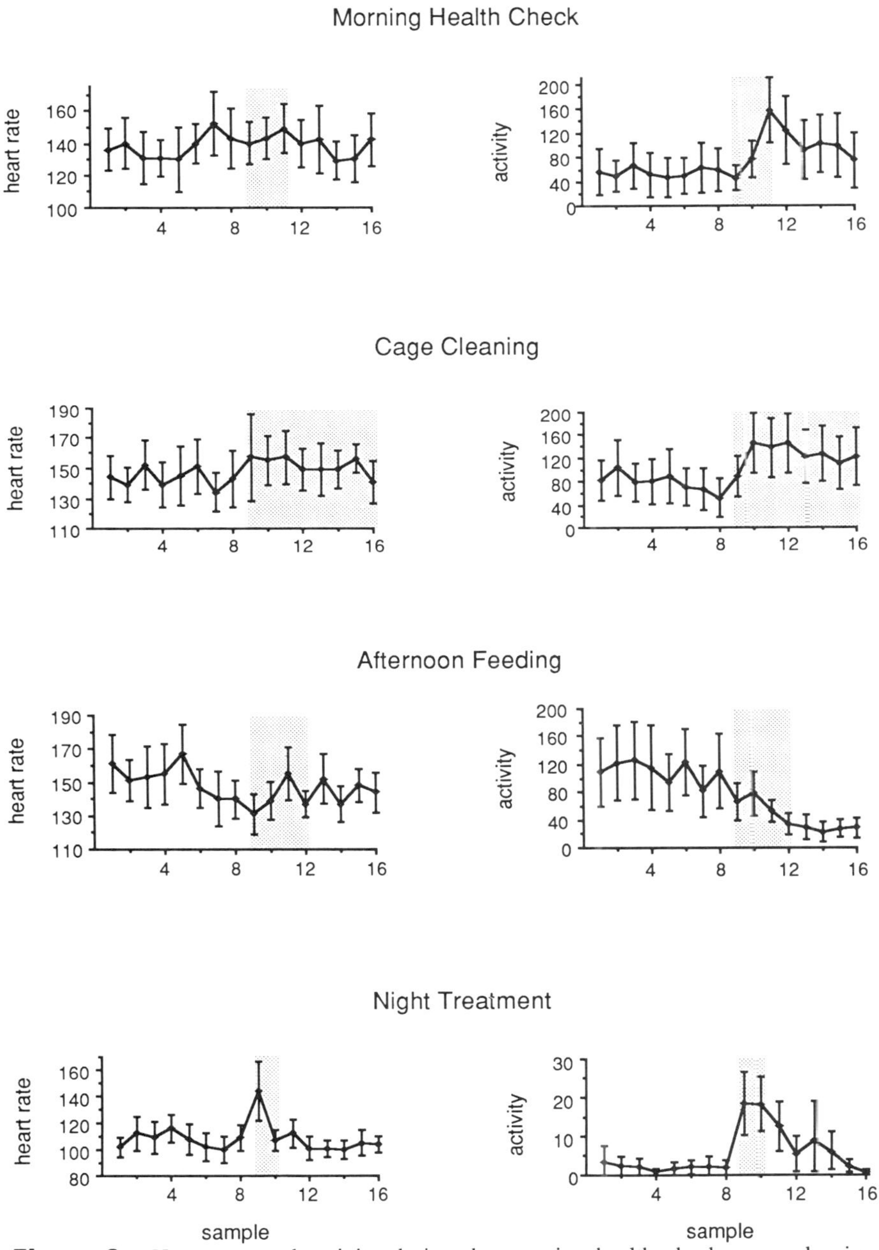

Figure 2 Heart rate and activity during the morning health check, cage cleaning, afternoon feeding, and night treatment. Each point represents the mean (± SEM) of samples taken on 4–9 different days. Samples were taken at 16 consecutive 2-minute intervals from female rhesus monkeys. The shaded bar denotes the period of time that a human was in the animal room.

In contrast to the transient effects of daily events, the reactions to less frequent procedures, such as tuberculin (TB) testing and cage changing, were larger in magnitude and longer lasting. Cage changes took place every 2 weeks and involved two or three technicians. The monkeys were placed in transfer boxes for about 15 minutes while their dirty cages were removed from the wall mounts and replaced with clean ones. Changing all of the cages in the room took about an hour and created high levels of noise. Mean heart rate during one cage change operation was 175 bpm, compared to 117 bpm recorded during that hour the following day (Figure 3). Activity levels, however, were not substantially different on the cage change day.

Peak sound levels during the procedure exceeded 100 decibels and may account for the observed increase in heart rate. Confinement in the transfer box probably also contributed to the elevated heart rate. In a previous study, 5 minutes of transfer-box confinement was associated with significantly increased plasma cortisol (Line, Clarke, & Markowitz, 1987). Once the cage change was completed and the technicians left the room, it was several hours before heart rates returned to the expected level.

The monkeys' response to TB testing was similar to their reaction to cage changing. TB testing was performed every 4–6 months and required that the subjects be tranquilized with ketamine. The heart-rate response to one TB test is presented in Figure 3. As expected, the monkeys' heart rates under the influence of ketamine were increased. At 1000 hours, while the monkeys were tranquilized, the mean heart rate was 192 bpm, in contrast to the usual mean heart rate of 131 bpm during that hour. The monkeys were fully awake an hour later, but heart rates did not return to expected levels for another 2 hours after that. Activity levels were also lower than usual for 2–3 hours following TB testing.

The changes in heart rate associated with human activity in the animal room may indicate that changes also occurred in other physiological systems, concurrently with these activities. Evidence to support this hypothesis was found in the results of a series of mild stress tests. The stressors used were procedures similar to those used in routine colony management. Heart rate alone was recorded during ten restraint tests (15 minutes of restraint using the cage-squeeze mechanism); heart rate and plasma cortisol were sampled in four additional restraint tests. Heart rate and plasma cortisol were also sampled before and 15 minutes after a 5-minute confinement in a transfer box attached to the front of the monkey's cage.

Blood samples (1 mL each) were collected from each subject by venipuncture into syringes containing sodium ethylenediaminetetraacetic acid (Na EDTA). All samples were taken between 1200 and 1530 hours in order to avoid variation due to circadian rhythms. A baseline sample was taken from the monkey in its home cage within 2 minutes of entry into the animal room. A second sample was taken immediately following the restraint or 15 minutes after release from the transfer box. Blood samples were analyzed for levels of plasma cortisol by

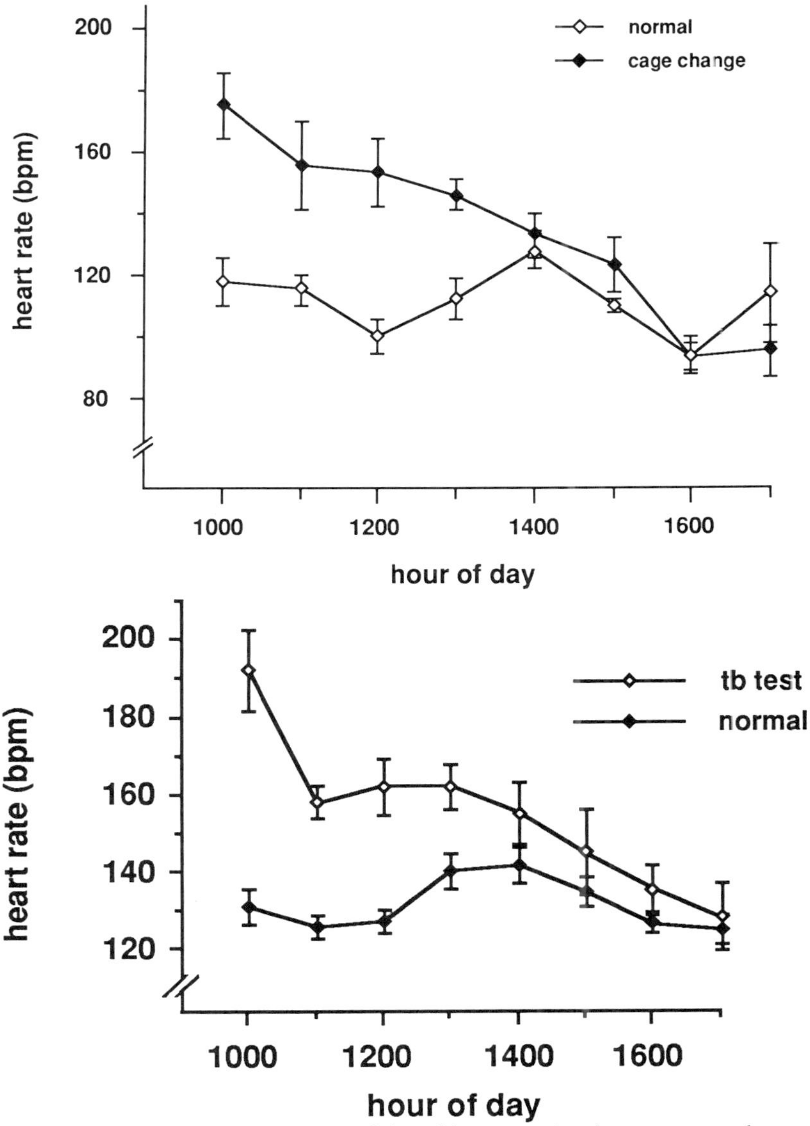

Figure 3 Upper graph: Heart rate of the subjects on a day the cages were changed and a normal day. The cage change took place between 0900 and 1100 hours. Each point represents the mean (± SEM) of 30 samples taken during a 1-hour period from the six monkeys. The normal curve is taken from data recorded the day after the cage change. Lower graph: Heart rate of the subjects on a day they were TB-tested and the average for a normal day. The monkeys were tranquilized with ketamine at 1000 hours and were fully awake by 1100 hours. Each TB test point represents the mean (± SEM) of 30 samples taken during a one-hour period for the six monkeys.

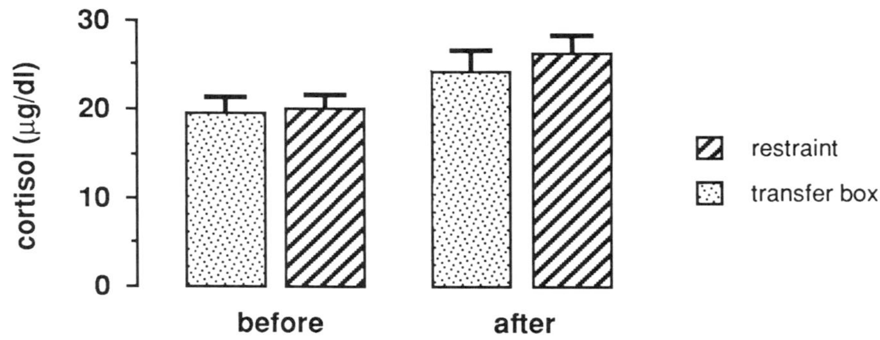

Figure 4 Changes in plasma cortisol in response to mild stressors. Each bar represents the mean (± SEM) for 10 adult female rhesus monkeys. Follow-up samples were taken immediately after 15 minutes of restraint in the home cage and 15 minutes after 5 minutes of confinement in a transfer box.

radioimmunoassay. Restraint with the cage-squeeze mechanism and confinement in a transfer box were both associated with significant increases in plasma cortisol (repeated measures ANOVA, $F = 13.47$, $p = .0001$ for restraint, and $F = 6.56$, $p = .002$ for transfer box, Figure 4). Heart rate also increased by 25 to 33 bpm during restraint tests, regardless of whether or not a blood sample was taken (2-way repeated measures ANOVA, $F = 11.4$, $p < .025$). Heart rate remained elevated for as long as 45 minutes following release from restraint.

We were not surprised that heart rate and activity patterns covaried with human activity in the animal room. Though the monkeys could communicate with one another visually and vocally, they were otherwise isolated and in a restricted environment over which they could exert little control. Except for lighting, all major environmental changes were directly connected with animal caretakers or research personnel. The monkeys usually watched humans who were in the room and directed their gaze to the room door when they heard the outer building door open. Heart rate and activity generally changed quickly when a human entered, but also returned to baseline levels soon after the person left the room.

We did not expect to find the prolonged changes in heart rate and activity associated with cage changing and TB testing. The monkeys had all had prior experience with these events, and we anticipated that they would have become habituated to them and would return to baseline heart rate or activity soon after the events ended (as they did in response to daily activities). Perhaps simply because these events lasted longer, the reaction to them also lasted longer. It is also possible that they were more aversive and thus generated longer lasting reactions. However, it may be that repeated exposure to an aversive procedure, no matter what its duration, is not sufficient to habituate animals to it or to otherwise diminish its impact. Repeated exposure to brief restraint did not lead to habituation of the heart-rate response. The magnitude of increase was just as large on the 12th test as on the first.

Perhaps the intensity of such reactions to aversive husbandry events could be reduced by making them more predictable or by actively training the animals to tolerate them. Most of them already occur on a regular schedule, though, so it is difficult to see how they could be made more predictable. Alternatively, the excitement caused by these activities may be beneficial by providing stimulation in an otherwise monotonous environment (see our conclusions below).

CAGE SIZE: MORE IS NOT NECESSARILY BETTER

Other workers have reported changes in maladaptive or otherwise abnormal behavioral responses when variables such as space and environmental richness are altered. Studies of cage size alone as a factor have produced conflicting results, however. In a comparison of three cage sizes, Draper and Bernstein (1963) found that stereotyped behavior was higher when rhesus monkeys were housed in small cages than when those same animals were housed in either of two larger cages. In a similar study, Paulk, Dienske, and Ribbens (1977) reported a slight decrease in abnormal locomotor behaviors in rhesus monkeys housed in a larger cage but no effect of cage size on nonlocomotor abnormal behavior. Goosen (1988) reported no change in abnormal behavior among singly caged rhesus macaques that were housed in cages six times larger than their standard cage.

Changes in cage size for nonhuman primates are mandated in the (1990) proposed United States Department of Agriculture regulations. Replacing all of the single cages currently in use would be very expensive. Estimates for the total cost of implementing the proposed regulations exceed $1 billion (Holden, 1988; see also T.L. Wolfle, chapter 13 in this volume). It would thus seem prudent to know whether increased cage space leads to demonstrable improvements in the animals housed in them before devoting large amounts of resources to buying new cages. Given the variability in results cited above, we felt it would be useful to assess further the effects of increased cage size on behavior and heart rate (Line et al., 1989).

Our subjects were the same six adult female rhesus macaques used to measure the effects of routine procedures. Telemetry was used to record heart rate and activity as described above. The monkeys were housed individually in stainless-steel squeeze-back cages of two different sizes (standard, 0.4 m^2 of floor space; and larger, 0.56 m^2 of floor space) in the same room. The standard cage corresponded to size requirements mandated by current regulations for Group 3 nonhuman primates (*Guide for the Care and Use of Laboratory Animals, 1985*). The larger cage could hold Group 4 primates and was 40%

larger in volume than the standard cage. Each subject was housed in the larger cage for one week at a time on four different occasions, using a counterbalanced schedule.

Direct behavioral observations were performed for 25 minutes three times a week by an observer seated across from the monkey's cage. All monkeys were well-habituated to this technique. A portable computer and bar code system were used to monitor frequency and duration of behavior. Frequency data were recorded for aggressive behavior (cage-shaking or open-mouth threats directed at the observer or other monkeys), submissive behavior (lipsmacking or fear-grimacing), cage manipulation (oral or manual exploration of any part of the cage), abnormal behavior (self-biting, self-holding, self-sucking, bizarre postures, plucking fur, head tossing, urine drinking, and saluting), and self-abusive behavior (a subset of abnormal behavior, i.e., self-biting and plucking fur). We also recorded the duration of looking at the observer, grooming, time spent in the front half of the cage, and nonstereotypic and stereotypic locomotion (circling or pacing). Interobserver reliability exceeded 0.90 for all behavioral categories.

Neither heart rate nor activity varied significantly between the two cage sizes (Figure 5). There was a significant variation with respect to time of day, as was found in the response to routine events. Among the behavioral measures recorded by direct observation in the two cage sizes, only vocalizations approached a significant difference (Friedman test, $\chi^2 = 13.99$, $p = .0514$). Analysis of the individual weeks revealed that vocalizations increased only during the first time the monkeys were moved to the larger cage (Wilcoxon test, $Z = 2.2$, $p = .0277$). This was the first time the subjects had been moved in 6 months, and it placed them in view of new neighbors. Since a similar increase did not occur the following three times in the larger cage, we concluded that the novel neighbors, not the size of the cage, were responsible for the change in behavior (Line et al., 1989).

Negative results should be interpreted cautiously. However, the fact that we consistently found that increased cage size did not change behavior or heart rate and the fact that previous reports did not consistently find a positive effect of increased cage size suggest that if the goal is to enhance psychological well-being, then spending large amounts of money to buy new cages may be a waste of resources. If money is to be spent to improve laboratory environments, then methods that are demonstrably effective should be given priority over ones that are intuitively appealing but lack experimental support. Increasing opportunities for socialization (see F.B.M. de Waal, chapter 9 in this volume & V. Reinhardt, chapter 16 in this volume), use of cage enrichment devices (Preilowski, Reger, & Engele, 1988; and see below), and providing increased foraging opportunities are techniques that have been shown to enhance the lives of laboratory primates.

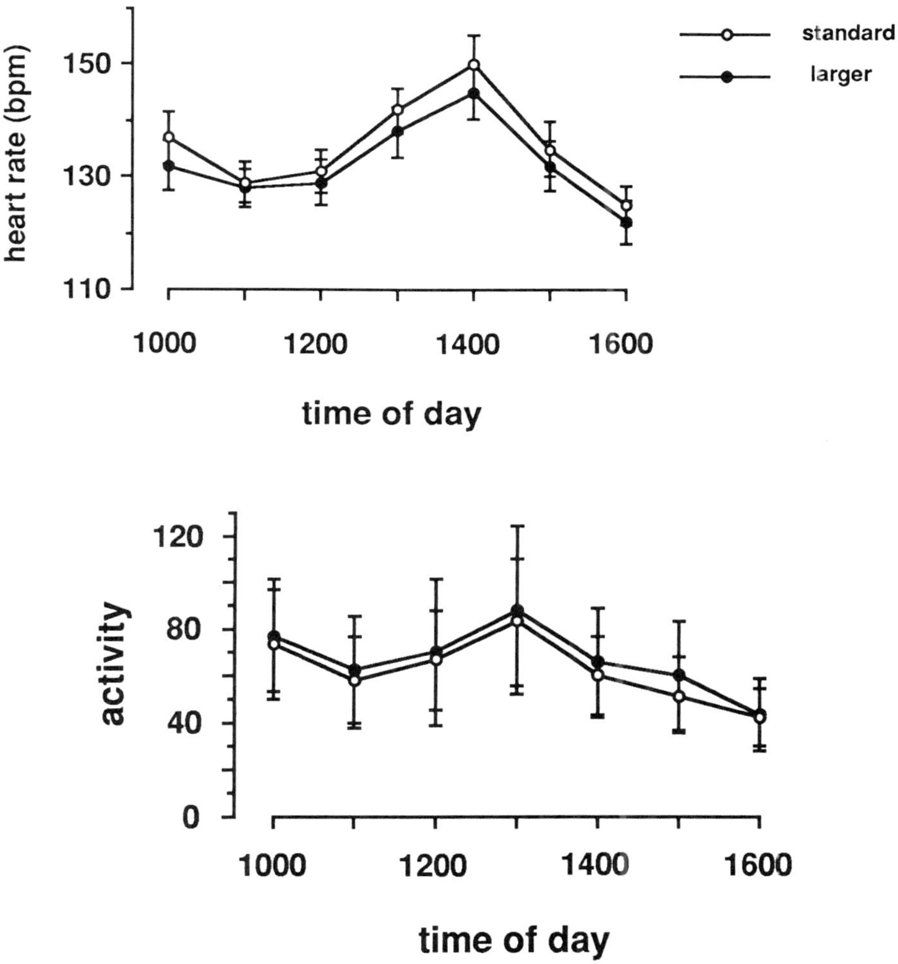

Figure 5 Changes in heart rate and activity in response to increased cage size. Each point represents the mean (± SEM) of 30 samples taken during a 1-hour period on 20 different days in standard and in larger cages. Data were recorded, via telemetry, from six adult female rhesus monkeys.

ENVIRONMENTAL ENRICHMENT: COMPLEXITY AND CONTROL

Another approach to improving laboratory environments has been to increase their complexity. Effects of providing animals with "furniture" and methods for self-feeding or manipulable objects in the form of toys, puzzles, branches, and sticks have been investigated (e.g., Champoux, Hempel, & Reinhardt, 1987; Gilbert &

Wrenshall, 1989; Markowitz, 1979, 1982; Reinhardt, Houser, Cowley, & Champoux 1987; Renquist and Judge, 1985; Ross & Everitt, 1988; for a summary of recent work see Line et al., 1989). Many of these studies, however, provide no baseline information with which to compare their results—a problem that makes interpretation of their data difficult, since past experience and previous behavioral repertoire of the animals cannot be eliminated as factors contributing to the variability in results obtained.

Monitoring behavior is one of the most commonly advocated methods for assessing psychological well-being. Yet there are very few published reports on the typical behavioral patterns expressed by singly caged primates in a laboratory environment. Without these baseline data, there is no rational basis for using behavioral measures to evaluate well-being in laboratory primates, nor for evaluating the results of changes in standard procedures. We have tested the effectiveness of several objects proposed to enhance well-being by comparing the behavioral and physiologic measures described above during alternative periods with and without the enrichment devices. In the remainder of this chapter, we describe the results of our most recent study of a device that provided music and food.

The apparatus was a battery-operated music-feeder box that allowed the monkeys to control some aspects of their immediate environment by simply touching one of three rods that extended a short distance into their cages. One rod turned a radio on and a second turned it off. The radio was tuned to a local "top 40" music station. The third rod operated a food dispenser and was independent of the radio control. Initially one 45-mg banana-flavored food pellet was delivered for five touches of the food dispenser rod. The ratio was increased by five responses per pellet each day that the monkey received 500 or more pellets. If the cumulative number of pellets delivered was less than 500 over 2 days, then the ratio was reduced to 5 responses per pellet.

The 10 subjects in this study included the same six adult female rhesus macaques used to measure the effects of routine procedures and cage size. Heart rate and activity were monitored via telemetry as described above. Behavioral observations were recorded three times a week on these animals (using the same technique described above), and we also used an additional four monkeys that were provided with the enrichment apparatus but that did not have telemetry implants. The monkeys were monitored for 12 weeks to obtain baseline data and were then given access to the music-feeder box for 12 weeks. The brief restraint test described above was performed seven times during each 12-week block (two tests in each block included blood sampling for cortisol determination before and after restraint). Additional blood samples for baseline plasma cortisol determination were taken two other times during each block.

The monkeys' rate of feeder use increased from roughly 2,200 responses per day the first week to 9,600 responses per day the next week and then remained

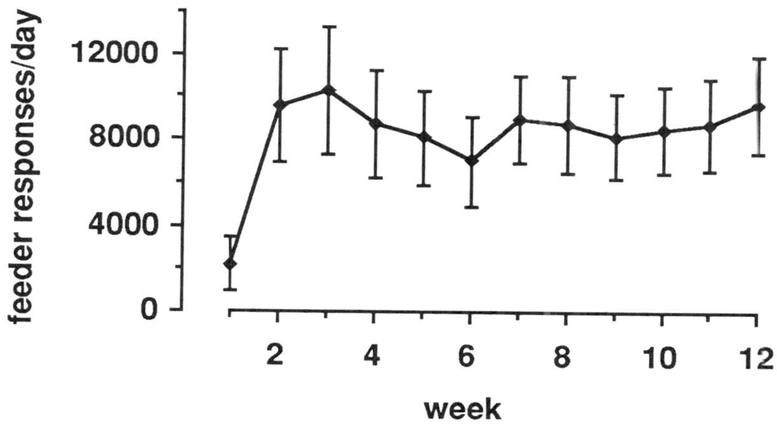

Figure 6 Use of the feeder control of the music-feeder box. Each point represents the mean (± SEM) number of responses per day for one week by all 10 subjects. Responses were comparatively low the first week, while the monkeys maintained a low pellet to response ratio, and then remained stable for the next 11 weeks.

at about that level throughout the remainder of the 12-week experimental period (Figure 6). Overall, the average number of touches to the feeder control rod was 8,221 per day, a response rate that earned an average of 562 pellets per day. The radio was also used throughout the 12 weeks it was available, with an average playing time of 76 minutes per day.

For the six animals with telemetry implants, there were no significant changes in activity associated with the music-feeder box. Overall heart rate during the daytime was increased by about six beats per minute in the presence of the apparatus. This effect was statistically significant (ANOVA, $F = 7.96$ $p < .05$) but was much smaller than the magnitude of variation in heart rate with time of day. There were no significant differences in nighttime heart rate (mean of all samples taken between 2400 and 0400, ANOVA, $F = 0.57$, $p > .1$).

Baseline plasma cortisol levels were determined from blood samples taken using the protocol outlined above. Four samples were taken during the seven months prior to introduction of the music-feeder box. Four additional samples were taken during the 3 months the enrichment device was available.

The mean basal plasma cortisol levels ranged from 24.41 ± 2.05 μg/dL to 30.4 ± 2.05 μg/dL during the baseline period, and from 15.9 ± 1.25 μg/dL to 20.1 ± 2.73 μg/dL during the period when the music-feeder box was present. The four samples during each period were averaged to obtain one control and one experimental basal cortisol level for each monkey. Comparison of these two means revealed a significant decrease associated with the presence of the enrichment apparatus (one-tailed t-test, $t = 5.79$, $p = .0002$).

Table 1

SIGNIFICANCE LEVELS FOR RESTRAINT TEST TWO-WAY REPEATED MEASURES ANOVAs, BEHAVIOR MONITORING STUDY

Variable	Condition: Box off, Box on		Time: Before, After		Interaction	
	F	*p*	*F*	*p*	*F*	*p*
Activity count	13.5	<.025	12.7	<.025	0.20	ns*
Heart rate	8.76	<.05	11.0	<.025	3.81	ns
Plasma cortisol	12.6	<.01	25.8	<.005	0.51	ns

Note. Sample sizes: Activity = 6, heart rate = 6, cortisol = 10. *Condition* compares tests with and without the music-feeder box in the cage. *Time* compares responses before and after restraint.
*not significant.

Several measures of behavior changed significantly when the music-feeder box was present. Cage manipulation decreased from a frequency of 11.9 bouts per 25-minute session to 7.2 bouts per session (Wilcoxon test, Z = 5.889, p = .0093). Abnormal behavior decreased from a frequency of 9.3 bouts per session to 4.5 bouts per session (Wilcoxon test, Z = 2.803, p = .0051). In addition, self-grooming decreased from 477 to 277 seconds per session (Wilcoxon test, Z = 2.803, p = .0051), time spent sitting decreased from 1,299 to 1,099 seconds per session (Wilcoxon test, Z = 2.803, p = .0093), and time standing increased from 94 to 251 seconds per session (Wilcoxon test, Z = 2.803, p = .0093). Nonstereotypic movement increased from 58 to 80 seconds per session, and stereotypic pacing declined from 60 to 48 seconds per session, although these differences were not statistically significant.

The monkeys spent an average of 360 seconds per session using the music-feeder box. The behavioral changes observed when the music-feeder box was present can be summarized as follows. Use of the box occupied about one fourth of the monkey's time; the frequency of abnormal behavior and cage manipulation decreased; the duration of grooming and sitting decreased; and the duration of standing increased.

Brief restraint tests that included blood sampling were conducted during the baseline period and while the music-feeder box was present. Activity and heart-rate values for the six monkeys with transmitters were averaged for 15 minutes before and after these tests. Plasma cortisol was measured immediately before and after restraint for all 10 subjects. The effects of the music-feeder box and restraint on these variables were examined using a repeated measures two-way analysis of variance. Significance levels for the analyses are presented in Table 1. Restraint increased activity, heart rate, and cortisol in all cases. Prerestraint heart rate and activity counts were higher during the tests in which the music-feeder box was present, and prerestraint cortisol values were lower. None of the interaction effects was significant.

In addition, heart-rate reactivity to restraint was assessed as a function of the presence of the enrichment device. The magnitude of increase in heart rate during restraint was compared with the change in heart rate following release from restraint. The increase was significantly less when the music-feeder box was present (2-way ANOVA, $F = 15.9$, $p < .025$). The heart rate also returned to baseline faster during this time.

Although individual response rates varied, all 10 monkeys used the music-feeder box throughout the 12 weeks it was available. In contrast to plastic toys or other simple objects, the apparatus produced no apparent novelty effect. Foraging is one of the major activities of free-ranging primates, so it is perhaps not surprising that they would maintain interest in a device that allows them to work for a food reward.

The overall increase in heart rate when the music-feeder box was present is a little more difficult to understand. It may have been a result of changes in posture (i.e., when the device was on their cage, animals stood upright more in order to use it) or in sound levels during those times that the radio was on. An alternative explanation is that the monkeys were more aroused while the enrichment device was present. Even though the effect was statistically significant, it was small in magnitude (the changes with time of day were much larger) and may have been physiologically unimportant.

Decreases in the frequency of cage manipulation and abnormal behavior while the music-feeder box was present suggest that use of the apparatus served as a substitute for these behaviors. Rhesus monkeys are naturally manipulative, and the enrichment device provided an attractive outlet for this sort of behavior. The decline in abnormal behavior is quantitative evidence of an increase in psychological well-being.

Besides the decrease in abnormal behavior, there were shifts in other behaviors toward levels displayed by free-ranging rhesus monkeys (Table 2). Autogrooming occupied almost a third of the observation time during the baseline period and was much lower when the music-feeder was added to the cage. It is unlikely that all of the grooming was required for skin care, and the decrease during the experimental period suggests that the initially high level could have been due to boredom.

Even though autogrooming rates declined while the box was present, they were still higher than the amount of autogrooming and allogrooming combined among free-ranging rhesus. Monkeys spent 24% of their time using the music-feeder box, a percentage similar to the amount of time devoted to foraging among free-ranging rhesus monkeys. The changes in sitting and standing were probably due to box use, since monkeys frequently stood in front of the apparatus while using it. The amount of time spent moving was lower in both conditions for the

Table 2
MEAN PERCENTAGE OF TIME DEVOTED TO VARIOUS BEHAVIORS BY SINGLY CAGED MONKEYS IN THIS STUDY AND BY FREE-RANGING MONKEYS

Source:	This Study		Malik, 1986	Seth & Seth, 1986			Marriott, 1988	
Condition:	Singly caged, Indoor		Free-Ranging, India				Free-Ranging, Cayo Santiago	Free-Ranging, Nepal
	Box Off	Box On	Fort	Urban	Roadside	Forest	Island	Forest
Behavior:								
Grooming	30.5[a]	17.7[a]	11.5	14.5	9.0	1.8	12.3	9.8
Locomotion	3.9	5.4	18.6	17.2	20.6	26.2	23.1	39.4
Foraging	—	—	8.0	16.2	23.3	40.1	10.8	15.6
Box Use	—	24.1	—	—	—	—	—	—

[a]autogrooming only

monkeys in this study than amounts observed in free-ranging monkeys (Malik, 1986; Marriott, 1988; Seth & Seth, 1986). Considering the restricted space available to the indoor monkeys, this is to be expected.

The lowered basal plasma cortisol (Figure 4) is an intriguing finding, but one that should be interpreted with caution. One possibility is that the unenhanced single cage is an aversive environment and that addition of the enrichment device led to the decrease in cortisol. The higher levels of cortisol measured in the control period may be analogous to the elevated basal levels seen in depressed humans (Ceulemans, Westenberg, & Van Traag, 1985). Another possibility is that underlying seasonal changes produced the pattern observed. Further testing will be required before any definitive conclusions can be drawn.

The lower heart-rate reactivity in response to restraint during the experimental period is consistent with the interpretation that addition of the music-feeder box enhanced the monkeys' ability to cope with stressful events. Even though the device afforded the monkeys control over only a limited part of the environment, they had control of virtually nothing prior to the presentation of the music-feeder box. These findings are consistent with earlier suggestions (Champoux, Schneider, & Suomi, 1987; Mineka, Gunnar, & Champoux, 1986) that ability to control environmental elements may have positive effects on socioemotional development.

The decrease in basal cortisol levels in heart-rate reactivity to restraint, in frequency of abnormal behaviors, and the changes in the pattern of normal behaviors indicate a positive effect of the enrichment device on the monkeys' psychological well-being. We suggest that the opportunity for the animals to control some limited aspects of their otherwise restricted environments by using the music-feeder box may be the key element in obtaining these results.

CONCLUSIONS

Baseline behavioral and physiologic data collected at the California Regional Primate Research Center show persistent and robust effects from human activities in the animal room, including responses to sporadic events as well as routine maintenance occurring at regular times each day. These observations should be considered in planning experimental treatments and in interpreting their results. Though we may assume that the monkeys are familiar with recurring procedures, they continue to react to them. We predicted that the monkeys would have been habituated to routine activities and that their reactions to them would be minimal or at least of short duration. The fact that this was not the case suggests that many of these events have a bigger impact on the animals than is generally recognized.

It would not be desirable, nor possible, to eliminate completely the types of human activities to which animals are apparently reactive. Even though some procedures (cage-changing for example) may be aversive to monkeys, these procedures are necessary to maintain healthy animals. Total removal of potentially arousing events may result in a monotonous environment that is also distressful. Relief of unnecessary distress is a reasonable goal, but determining specifically where to draw the line between tolerable and intolerable distress is going to be difficult. Experience with mildly stressful events may be beneficial, by providing monkeys with experience in coping. This experience could better enable them to contend with other events that are unavoidable, much as early experience with mild stressors allows rodents to tolerate later aversive situations better than animals that were sheltered from outside stimulation as neonates (Denenberg & Morton, 1962; Weininger, 1956).

These baseline observations also illustrate the type of background data necessary for empirical assessment of the effectiveness of various proposals to enhance psychological well-being. Without this sort of baseline, systematic evaluation is simply not possible. Most of us will agree that it is important to provide richer environments for animals that serve as research subjects. However, it will not be sufficient simply to make changes that *seem* to us as though they should provide the animals with some improvements over previous confinement conditions. The very real danger that what is intuitively obvious may, in fact, be incapable of producing the desired effect is demonstrated by the failure to obtain significant effects on behavioral or physiological variables by increasing cage size in this as well as in other studies.

Preliminary data from similar systematic inquiry indicate that affording the monkeys some degree of control in an otherwise extremely limited environment, through provision of devices such as the music-feeder box, will be more effective. In order that we may ultimately be cost-effective in our efforts and simultaneously provide *measurable* improvements in the lives of our subjects, intensive scientific evaluation of various enrichment strategies is necessary at the outset.

References

Ceulemans, D. L., Westenberg, H. G., & Van Traag, H. M. (1985). The effect of stress on the dexamethasone suppression test. *Psychiatric Research, 14*, 189–195.

Champoux, M., Hempel, M., & Reinhardt, V. (1987). Environmental enrichment with sticks for singly-caged adult rhesus monkeys. *Laboratory Primate Newsletter, 26*, 5–7.

Champoux, M., Schneider, M., & Suomi, S. (1987). Environmental enrichment and affective responses in infant rhesus monkeys. *American Journal of Primatology, 12*, 335.

Clarke, A. S., Line, S. W., Ellman, G., & Markowitz, H. (1987). Hormonal and behavioral responses of rhesus macaques to an environmental enrichment apparatus. *American Journal of Primatology, 12*, 335–336.

Denenberg, V. H., & Morton, J. R. C. (1962). Effects of environmental complexity and social groupings upon modification of emotional behavior. *Journal of Comparative and Physiological Psychology, 55*, 242–246.

Draper, W. A., & Bernstein, I. S. (1963). Stereotyped behavior and cage size. *Perceptual and Motor Skills, 16*, 231–234.

Gilbert, S. G., & Wrenshall, E. (1989). Environmental enrichment for monkeys used in behavioral toxicology studies. In E. Segal (Ed.), *Housing, care and psychological well-being of nonhuman primates* (pp. 244–254). Park Ridge, NJ: Noyes.

Goosen, C. (1988). Developing housing facilities for rhesus monkeys: Prevention of abnormal behavior. In A. C. Beynen & H. A. Solleveld (Eds.), *New developments in biosciences: Their implications for laboratory animal science* (pp. 67–70). Dordrecht: Martinus Nijhoff Publishers.

Hediger, H. (1964). *Wild animals in captivity*. New York: Dover Publications.

Holden, C. (1988). Billion dollar price tag for new animal rules. *Science, 242*, 662–663.

Line, S. W. (1987). Environmental enrichment for laboratory primates. *Journal of the American Veterinary Medical Association, 190*, 854–859.

Line, S. W., Clarke, A. S., Ellman, G., & Markowitz, H. (1987). Behavioral and physiologic responses of rhesus macaques to an environmental enrichment device. *Laboratory Animal Science, 37*, 509.

Line, S. W., Clarke, A. S., & Markowitz, H. (1987). Plasma cortisol of female rhesus monkeys in response to acute restraint. *Laboratory Primate Newsletter, 26*, 1–4.

Line, S. W., Clarke, A. S., Markowitz, H., & Ellman, G. (1990). Responses of female rhesus macaques to an environmental enrichment apparatus. *Laboratory Animals, 24*, 213–220.

Line, S. W., Morgan, K. N., Markowitz, H., & Strong, S. (1989a). Influence of cage size on heart rate and behavior in rhesus monkeys. *American Journal of Veterinary Research, 50*, 1523–1526.

Line, S. W., Markowitz, H., Morgan, K. N., & Strong, S. (1989b). Evaluation of attempts to enrich the environment of singly-caged nonhuman primates. In J. Driscoll (Ed.), *Animal care and use in behavioral research: Regulations, issues and applications*, (pp. 103–117). Washington, DC: Animal Welfare Information Center, National Agriculture Library.

Line, S. W., Morgan, K. N., Markowitz, H., & Strong, S. (1989). Heart rate and activity of rhesus monkeys in response to routine events. *Laboratory Primate Newsletter, 28*, 1–4.

Malik, I. (1986). Time budgets and activity patterns in free-ranging rhesus monkeys. In J. G. Else & P. C. Lee (Eds.), *Primate ecology and conservation* (pp. 105–114). Cambridge: Cambridge University Press.

Markowitz, H. (1979). Environmental enrichment and behavioral engineering for captive primates. In J. Erwin, T. Maple, & G. Mitchell (Eds.), *Captivity and behavior* (pp. 217–238). New York: Van Nostrand Reinhold.

Markowitz, H. (1982). *Behavioral enrichment in the zoo*. New York: Van Nostrand Reinhold.

Markowitz, H., & Line, S. W. (1989). Primate research models and environmental enrichment. In E. Segal (Ed.), *Housing, care and psychological well-being of nonhuman primates* (pp. 203–212). Park Ridge, NJ: Noyes.

Marriott, B. M. (1988). Time budgets of rhesus monkeys (*Macaca mulatta*) in a forest habitat in Nepal and on Cayo Santiago. In J. E. Fa and C. H. Southwick (Eds.), *Ecology and behavior of food-enhanced primate groups* (pp. 125–149). New York: Alan R. Liss.

Mineka, S., Gunnar, M., & Champoux, M. (1986). Control and early socioemotional development: Infant rhesus monkeys reared on controllable versus uncontrollable environments. *Child Development, 57*, 1241–1256.

National Research Council. (1985). *Guide for the care and use of laboratory animals*. (NIH Pub. No. 85-23). Washington, DC: U.S. Department of Health and Human Services.

Novak, M. A., & Suomi, S. J. (1988). Psychological well-being of primates in captivity. *American Psychologist, 43*, 765–773.

Paulk, H. H., Dienske, H., & Ribbens, L. G. (1977). Abnormal behavior in relation to cage size in rhesus monkeys. *Journal of Abnormal Psychology, 86*, 87–92.

Preilowski, B., Reger, M., & Engele, H. (1988). Combining scientific experimentation with conventional housing: A pilot study with rhesus monkeys. *American Journal of Primatology, 14*, 223–234.

Reinhardt, V., Houser, W. D., Cowley, D., & Champoux, M. (1987). Preliminary comments on environmental enrichment with branches for individually caged rhesus monkeys. *Laboratory Primate Newsletter, 26*, 1–3.

Renquist, D. M., & Judge, F. J. (1985). Use of nylon balls as behavioral modifiers for caged primates. *Laboratory Primate Newsletter, 24*, 4.

Ross, P. W., & Everitt, J. I. (1988). A nylon ball device for primate environmental enrichment. *Laboratory Animal Science, 38*, 481–482.

Seth, P. K., & Seth, S. (1986). Ecology and behavior of rhesus monkeys in India. In J. G. Else & P. C. Lee (Eds.), *Primate ecology and conservation* (pp. 89–103). Cambridge: Cambridge University Press.

U.S. Department of Agriculture, Animal and Plant Health Inspection Service, Animal Welfare; Standards; Proposed Rule (9 C.F.R., Part 3) [Docket No. 90-040]. Federal Register 55 (158), 33448-33531; Aug. 15, 1990.

Weininger, O. (1956). The effects of early experience on behavior and growth characteristics. *Journal of Comparative and Physiological Psychology, 49*, 1–9.

CHAPTER 18

APPLIED PRIMATE ECOLOGY: EVALUATION OF ENVIRONMENTAL CHANGES INTENDED TO PROMOTE PSYCHOLOGICAL WELL-BEING

JOSEPH M. ERWIN

We have been asked to consider ways in which the "psychological well-being" of nonhuman primates confined in laboratories can be enhanced. Here I present some thoughts and evidence regarding the relationships between physical and social environments and the behavior and health of primates. The evidence reviewed here comes from studies conducted by myself and colleagues in several environments, including natural settings, breeding colonies, research centers, and laboratories.

This document is not intended to serve as a comprehensive review of studies on the topics addressed; recent reviews are available (Line, 1987; Novak & Suomi, 1988). Our research results are presented to emphasize that common sense is not a reliable basis for primate husbandry decisions. The results of some of our studies

Environmental enrichment research was supported by NSF-SBIR grant ISI-8760708 to John Landon, SEMA, Inc. (J. Erwin, Consulting Primatologist). NIH Contracts: NO1-A152556, Pt. A, & NO1-CP-1079 also provided support. Earlier studies were supported by USPHS/NIH grants: RR00166 and HD08633. John Landon, J. Phillips, T. Moskal, K. Weld, B. Metz, and R. Nadler (Yerkes RPRC) were very helpful at SEMA, Inc., as were G. P. Sackett, O. Smith, J. Blakely, C. Crockett, and Nancy Erwin at the University of Washington.

were both unexpected and of great practical significance for colony management and primate health. Preliminary results are also presented from some experiments designed to evaluate techniques intended to promote the psychological well-being of individually caged primates of several species.

SOCIAL GROUPS

In 1974, I embarked on postdoctoral research at the Regional Primate Research Center at the University of Washington, at the breeding colony at Medical Lake. Before my arrival there, the housing configuration at the breeding station had been changed. Groups of pigtailed macaques (*macaca nemestrina*) that had been housed in single rooms (73.4 square feet of floor space) were given two adjoining rooms (146.8 square feet of floor space). The new configuration was adopted throughout the colony of over 1,000 primates specifically to reduce possible stress from crowding. A high rate of enteritis in the colony was thought to be related to stress (as well as to the cleaning procedure that involved sequential access of different groups to the same rooms).

When I arrived, it was quite clear that fighting and injuries were common occurrences. I had been advocating social housing of primates, but had not anticipated the potential consequences. I was asked by the director to conduct systematic behavioral observations to determine the causes of violence within groups and to suggest solutions to the problem. My first hypothesis was that fighting was associated with short-term crowding. The cleaning procedure at the time involved crowding each group into one room of a two-room suite while the empty room was cleaned. I suspected that most of the aggression and trauma occurred during the crowded period.

Study 1: Short-term Crowding

The first study involved observation of six groups of pigtailed macaques (5 adult males, 67 adult females, and 26 infants). Each group was observed while it had access to one room (an average of 4.5 square feet per animal) and then to two rooms (an average of 9 square feet per animal). Two observation sequences were employed: observation first in one room, then in two rooms, and finally in one room (hereafter referred to as the 1-2-1 sequence), or first in two rooms, then in one room, and finally in two rooms (the 2-1-2 sequence). Female-female fights accounted for most of the observed aggression, and remarkably, more contact aggression and threatened aggression occurred when the animals had access to both rooms than when they had access to only one room. These results were consistent across all phases of the study and across all groups but one—an all-female group whose responses were mixed. We simply could not believe that aggression was inversely related to the amount of physical space available, so we designed another study to further evaluate the effects of short-term spatial crowding.

Study 2: Short-term Crowding Replication

The second study involved observations of eight groups of pigtailed macaques, none of which had been involved in the previous study (8 adult males, 79 adult females, and 22 infants). Each group included one adult male. The average amount of space available per animal was 5.4 square feet in the single room and 10.8 square feet when both rooms were accessible. Each group was observed in 1-2-1 and 2-1-2 sequences, but two control sequences were added (1-1-1, 2-2-2). The results were even more dramatic than in the first study. Without exception, groups exhibited more contact and threatened aggression when they had access to two rooms (Anderson, Erwin, Flynn, Lewis, & Erwin, 1977; Erwin, 1977, 1979, 1986). The results were completely opposite to our expectation but were clearly consistent across all one-male groups.

Study 3: Effects of Group Size

Experimental studies of crowding are conducted in two basic ways. One involves systematic variation in the amount of space available (referred to as studies of *spatial density*); the other involves variation in the number of individuals within a specified amount of space (referred to as studies of *social density*). One kind of social density study simply compares the behavior of groups of different sizes. A post hoc analysis of the relationship between frequency of female-female contact aggression and the number of females per group in studies 1 and 2 revealed a significant correlation (Spearman's rho = .62; $p < .02$). This finding suggested a third study in which the behavior of all available groups was studied while groups were confined in single rooms. Of 62 one-male groups, 52 included between 8 and 13 females. At least six groups were available for each group size (N = 540 females in 52 groups). As before, the frequency of contact aggression among females was directly related to the number of females per group (Spearman's rho = .99). The relationship was just as strong when the data were adjusted to control for the effects of interaction potential (the probability of increased interaction based on chance alone as a function of increased numbers of interactors) (Erwin, 1979; Erwin & Erwin, 1976).

Studies of Group Composition Effects

Study 3 also yielded interesting results regarding effects of the number of males per group. Of the 92 groups of pigtailed macaques observed during this study, 13 contained no males, 62 contained one male, 13 contained 2 males, and 4 contained more than 2 males. The frequency of female-female aggression was much higher for groups containing no male than in those that included at least one male (Erwin, 1979). This result was consistent with other surveys we conducted to compare aggressive behavior in "harem" groups and all-female groups.

The first survey included six groups (N = 67 animals), three with no adult male and three with one male each. Each of two observers conducted separate observations of all groups, and each was completely blind to the data of the other. Far more aggression occurred in each all-female group than in any group containing a male (Erwin, 1979; Sackett, Oswald, & Erwin, 1975). A larger survey, involving 203 pigtailed macaques in 20 groups (15 adult males, 166 adult females, and 22 infants), compared the behavior of 15 one-male groups with five all-female groups. In 10 of the groups, there were no infants and in 10, there was at least one infant. Again, a blind replication was conducted. Females exhibited considerably less aggression in the presence of adult males than they did in all-female groups (Dazey, Kuyk, Oswald, Martenson, & Erwin, 1977). All these surveys suggested that the mere presence of an adult male inhibited aggression among females.

The Control Role of Males

If the presence of a male inhibits aggression among females, removal of males from groups should result in increased aggression among females, and reintroduction of males should reduce female aggression. We briefly removed males from six groups of pigtailed macaques (including 6 adult males, 51 adult females, and 7 infants), and the hypothesis was dramatically supported (Oswald & Erwin, 1976). In every group, female-female aggression increased at least fivefold when males were removed and returned to baseline levels when males were reinstated. A comparison study involving experimental removal of high-, medium-, or low-ranking females showed no effect from removing females (Swenson & Bartlett, 1976).

Direct observation of groups housed in two-room suites revealed that most female-female aggression occurred in the room where the male was not present. Female-female aggression near the male was usually limited to a single aggressive act, with the perpetrating female retreating into close proximity to the male. Retaliatory aggression toward females near males was never observed. In rooms away from the male, female-female aggressive episodes occurred as extended bouts of reciprocal fighting or repeated attacks directed toward one female by another.

The two-room suites constituted a high-risk environment, even though animals had access to twice as much space as in the single-room situation. The risk was largely based on the social organization and behavior of this species, although several other primate species are known to exhibit similar behavior. The disruption of the male "control role" probably accounted for most of the unexpected increase in aggression in two-room suites relative to single rooms; but females in all-female groups did not exhibit less aggression when they had access to two rooms than when they had access to only one. This result suggests that even in the absence of the male, variation in size of physical space had little influence on aggressive behavior.

The discovery that more aggression occurred when groups had access to two rooms than when they had access to only one resulted in another colony-wide change. Each group was assigned a two-room suite, but now only one room was accessible each day. A detailed follow-up of colony records showed that treatments for trauma and enteritis decreased markedly when available space was reduced to one room at a time. The effect was sustained across years (Erwin & Sackett, 1990).

INDIVIDUAL CAGING

I do not advocate individual caging of primates when better alternatives exist; however, there are many situations in which individual caging, at least for limited periods, is advisable or essential. Quarantine is necessary during outbreaks of communicable disease, when animals are being transported from one facility to another (especially for international shipments), during some clinical treatments, and during some vaccine-testing or other infectious disease studies. Individual caging is often (but not always) the best alternative for experimental situations requiring frequent access to primates, individual dietary control, postoperative recovery, metabolism studies, pregnant females at risk for clinical problems, studies involving indwelling catheters or neurological appliances, and some kinds of cognitive and psychopharmacological testing. Housing animals involved in such studies with conspecifics during periods when that is possible is a worthy goal that challenges us to be creative and practical at the same time.

Another approach to psychological well-being is to attempt to enrich the environments of individually housed primates in ways that provide opportunities for expression of species-typical behavior patterns, environmental control, and sensory and cognitive stimulation. The proposed standards under the Animal Welfare Act address this issue by requiring environmental enrichment, such as provision of manipulable objects to individually caged primates. Several studies have been conducted attempting to document or evaluate the usefulness of objects such as pet toys to promote primate psychological well-being. Early in 1988, I began serving as a consultant for SEMA, Inc. on a project funded by a grant from the National Science Foundation (NSF) to devise and evaluate techniques of providing environmental enrichment to individually caged primates, especially macaques. Later in 1988, I accepted a position as director of a new department of primate ecology at Diagnon Corporation, the parent company of SEMA. My primary duties involve development of an environmental enrichment program for SEMA.

One aspect of the NSF project involved evaluation of several commercially available pet toys as possible enrichment devices. The task was to determine which toys (or design features) were most effective in eliciting manipulation, promoting appropriate activity rates, and reducing the incidence of abnormal patterns of behavior. Kathy Weld and Becky Metz, the behavioral technicians for SEMA, have conducted a series of studies (cf. Weld, Metz, & Erwin, 1989) on the effectiveness of pet toys and other objects as enrichment devices. Although the detailed statistical analyses of all of these studies are not yet complete, some remarkable results are already apparent. The complete analyses will be published elsewhere, but some results are summarized below.

The effect of short-term exposure to a variety of pet toys was evaluated for chimpanzees (*Pan troglodytes*), rhesus macaques (*Macaca mulatta*), longtailed macaques (*Macaca fascicularis*), patas monkeys (*Erythrocebus patas*), and vervets (*Cercopithecus aethiops*). Species differences were very pronounced, with macaques and chimpanzees manipulating objects the most and patas and vervets manipulating them hardly at all. Age and gender differences in response to manipulable objects were also very clear for the longtailed macaques. The presence of any object virtually eliminated self-directed aggressive behavior for macaques and did so to a large extent for chimpanzees. Many of the macaques and chimpanzees continued to carry, manipulate, or play with the objects over a period of several months.

Chimpanzees often increased their use of particular objects over time and used objects and combinations of objects in new ways, months after their first exposure. As expected, the responses of chimpanzees to objects were highly variable.

The most effective objects have been adopted for regular use with individually caged chimpanzees and macaques. In one study of longtailed macaques, self-directed aggressive behavior was suppressed (relative to baseline) in the presence of a manipulable object for more than five months, for four out of five animals. Current studies include evaluation of long-term exposure to multiple objects, with or without replacement.

The chimpanzees at SEMA receive special attention. They have been involved in assessment of a spaced feeding schedule, responses to hand-held and wall mirrors, feeding of a variety of foods (e.g., corn-on-the-cob, coconuts, etc.), and exposure to many kinds of manipulable objects and toys. All chimpanzees at SEMA are captive born, and nearly all have been nursery reared. They typically arrive at SEMA late in the second year of infancy and are returned to the breeding colony of origin after 2–3 years. During the time they are at SEMA, they undergo a series of vaccinations and vaccine challenges, some of which require absolute quarantine. The quarantine units contain cages with more than 8.0 square feet of floor space. It has been widely alleged that pairs of chimpanzees are kept in isolators at SEMA that afford each animal about 1/16 of the cubic space that is actually available. Although the cages are not spacious, and I do not recommend or advocate keeping chimpanzees in cages of this size, the size is in full compliance with existing regulations, and the facility is fully accredited by the American Association for Accreditation of Laboratory Animal Care (AAALAC).

The quarantine units are equipped with lights and side windows that allow all chimpanzees visual and auditory access to their neighbors. Sound levels in the quarantine units have been measured and found acceptable for humans, and I can attest from sitting inside the chimpanzee cages that the high-pitched sound of the air handling system is inaudible. A regular series of seminars for staff members has been initiated dealing with topics in primate behavior and health. The SEMA facility is undergoing renovation that will provide the degree of biological containment, control, and access essential for infectious disease trials while offering expanded space, complexity, and social opportunities for chimpanzees.

Writing regulations is a difficult and thankless job. Those involved in the process have been remarkably patient with various factions and special interest groups—perhaps too patient with those organizations that have little or no direct expertise in the care and management of confined primates and whose members misrepresent themselves as spokespeople for the American public.

In my opinion, the regulations as originally proposed by the Animal and Plant Health Inspection Service included many unproductive and unwarranted requirements that contributed nothing to the well-being of primates. For example, the proposed humidity range did not reflect values experienced in the wild by primates. No definition was provided for the critical term *psychological well-being*. The definition of field work excluded virtually all useful field work and placed impossible burdens on field researchers.

I have been advocating what I consider to be a more productive approach—the preparation by each facility of a document that might be called a Primate Environmental Plan or a Plan for Primate Psychological Well-being. The document could provide a brief review of relevant scientific literature on the behavior and ecology of each species maintained, along with a specific description of the conditions of confinement and care. As techniques of environmental enrichment are scientifically validated, the plan could be revised to include implementation of the most appropriate and effective procedures. Fortunately, the revised proposed rule of August 15, 1990 has adopted this approach (USDA, 1990).

Efforts concerning the care, use, conservation, and understanding of primates are too important and too valuable to be destroyed by those who advocate the elimination of all animal research. Let us redouble our efforts to improve primate housing and husbandry progressively and to understand the relationships between primates and their environments.

Many widely held assumptions about primate responses to their environments have been falsified by careful scientific research. Studies of the effects of the quantity of available space on behavior and health clearly indicate that cage size per se is one of the least important factors in the promotion of primate well-being. The quality of the environment appears to be far more significant than the quantity of available space. Before we spend a billion dollars of public monies buying new cages that are only slightly larger than the old ones, we should clearly determine optimal and minimal cage sizes by conducting the needed research. From the current evidence, however, it appears that mandated increases in minimum cage size would be the most costly and least effective method of enriching the lives of confined primates.

References

Anderson, B., Erwin, N., Flynn, D., Lewis, L., & Erwin, J. (1977). Effects of short-term crowding on aggression in captive groups of pigtailed monkeys (*Macaca nemestrina*). *Aggressive Behavior*, *3*, 33–46.

Animal Welfare Act, 7 U.S.C. §§ 2131-2157 (1985).

Dazey, J., Kuyk, K., Oswald, M., Martenson, J., & Erwin, J. (1977). Effects of group composition on agonistic behavior of captive pigtailed macaques (*Macaca nemestrina*). *American Journal of Physical Anthropology*, *46*, 73–76.

Erwin, J. (1977). Factors influencing aggressive behavior and risk of trauma in the pigtail macaque (*Macaca nemestrina*). *Laboratory Animal Science*, *27*, 541–547.

Erwin, J. (1979). Aggression in captive macaques: interaction of social and spatial factors. In J. Erwin, T. Maple, and G. Mitchell (Eds.), *Captivity and behavior: Primates in breeding colonies, laboratories, and zoos*. (pp. 139–171). New York: Van Nostrand Reinhold.

Erwin, J. (1986). Environments for the captive propagation of primates: interaction of social and physical factors. In K. Benirschke (Ed.), *Primates: The road to self-sustaining populations* (pp. 297–305). New York: Springer Verlag.

Erwin, N., & Erwin, J. (1976). Social density and aggression in captive groups of pigtail monkeys (*Macaca nemestrina*). *Applied Animal Ethology*, *2*, 265–269.

Erwin, J., & Sackett, G. (1990). Effects of management methods, social organization, and physical space on primate behavior and health. *American Journal of Primatology*, *20*, 23–30.

Line, S. (1987). Environmental enrichment for laboratory primates. *Journal of the American Veterinary Medical Association*, *190*, 854–859.

Novak, M. A., & Suomi, S. J. (1988). Psychological well-being of primates in captivity. *American Psychologist, 43*, 765–773.

Oswald, M., & Erwin, J. (1976). Control of intragroup aggression by male pigtail monkeys (*Macaca nemestrina*). *Nature, 262*, 686–688.

Sackett, D., Oswald, M., & Erwin, J. (1975). Aggression among captive female pigtail monkeys in all-female and harem groups. *Journal of Biological Psychology, 17*, 17–20.

Sackett, G., & Erwin, J. (1981). Effects of weaning age on mortality of colony-born pigtailed macaques (*Macaca nemestrina*). *Theriogeneology, 15*, 505–511.

Swenson, L., & Bartlett, L. (1976). *Effects of brief removal of females from groups of pigtailed macaques* (*Macaca nemestrina*). Paper presented at the Congress of the International Primatological Society, Cambridge, United Kingdom.

U.S. Dept. of Agriculture, Animal and Plant Health Inspection Service, Animal Welfare; Standards; Proposed Rule (9 C.F.R., Part 3) [Docket No. 90-040]. Federal Register 55 (158), 33448–33531; Aug. 15, 1990.

Weld, K., Metz, B., & Erwin, J. (1989). Experimental evaluation of environmental enrichment techniques: provision of manipulable objects to five primate species. *American Journal of Primatology, 18*, 169.

PART SIX

PUBLIC REACTION TO PRIMATE RESEARCH

CHAPTER 19

TRANSLATING RESEARCH INTO WORKABLE REGULATIONS

B. TAYLOR BENNETT

This topic is included here because the animal rights issue has become a public policy issue, as exemplified by the inclusion of current language in the Amended Animal Welfare Act (PL-99-198) concerning the psychological well-being of primates. If research results are to be translated into workable regulations, we must convince the public and our legislative representatives that we are and have been concerned with animal welfare issues and that we care about the well-being of the animals in our charge.

The need for our increased participation in the public arena was clearly demonstrated at a recent legislative conference sponsored by the National Association for Biomedical Research (NABR). The senator responsible for inclusion of the language on psychological well-being of primates in the Animal Welfare Act stated that he had visited a laboratory in his home state that housed chimps and felt that they needed more space. He also indicated that all we needed to address this issue was more space and more money. When knowledgeable members of the audience tried to convince him that the issue was not simply a matter of space and money, he implied that the scientific community had no say-so in this matter.

Unfortunately, his response shows how, in many cases, the general public would react in a similar situation. This response is largely a result of the fact that the various animal rights groups have cultivated the public's natural allegiance to nonhuman primates; an allegiance that has interjected itself into the legislative process. At times, the public appears to assume that if nonhuman primates are good animal models because they are our closest living relatives, then they must have similar to our own, if not identical, needs.

Regardless of how we got where we are today, the fact remains that the Secretary of Agriculture has been given a congressional mandate to develop standards that address a particular characteristic, psychological well-being, for which few descriptive data are available. Hence, standards will, by necessity, be developed from input that may often be based more on subjective opinions than on the results of carefully evaluated scientific studies. In fact, based on my limited but expanding scientific knowledge of primate behavior, I am not sure that there is complete agreement among primatologists and primate behaviorists on many aspects of this complicated issue.

It would appear then that we are stuck on the proverbial horns of a dilemma. In short, we have a situation that must be addressed with some urgency. I use the word "we" to include: the research community, which will continue to rely on the use of nonhuman primates in studies where these animals represent the most appropriate model; the staff of the regulatory agency that must develop the new standards; the laboratory animal veterinarians who will bear the ultimate day-to-day responsibility for implementing them; and the primatologists and primate behaviorists who must contribute freely of their expertise and their time to the ongoing regulatory process. We must all work together to address the concerns of the public, as expressed in the congressional mandate of the Animal Welfare Act (1985).

As director of a large institutional animal-care program, I approach perceived problems in a different manner depending upon which member of the staff has approached me for advice. If one of the postdoctoral trainees perceives that he or she has a problem, the solution to it becomes what I like to call a "significant learning experience." This exercise is a three-step process:

1. Define the perceived problem.
2. Evaluate the operative facts that make the current situation a problem. (At this point it may be determined that what we really have is an original assessment of the situation by an inexperienced individual who saw a relatively routine challenge as a problem.)
3. Develop and implement a solution to include a follow-up evaluation.

Put simply, the process asks: a) what? b) why? and c) what are we going to do about it?

Now if a member of the senior staff approaches me under similar circumstances, the solution process may cease to be a learning experience. It may, in fact, evolve into an explanation of exactly what the person was thinking of at the time this problem began to occur and an explanation as to how it could have gotten this far without recognition.

The situation that faces us in respect to the amended Animal Welfare Act is probably a hybrid of the above. It certainly presents us with a significant learning experience, but one which might not be so challenging had we been more actively involved with the legislative process through communicating with the public and directly with our legislative representatives. So let's apply the three-step process to the current situation involving the amended Animal Welfare Act (1985).

WHAT?

On December 23, 1985, PL 99-198 was signed into law. Section 12 empowers the Secretary of Agriculture to promulgate humane standards for a physical environment adequate to promote the psychological well-being of primates. This requirement creates, through a lack of specificity, several issues that will have to be addressed.

1. What is meant by physical environment?
2. Will the standards deal solely with the physical environment, or will they take into account the many factors that go into addressing an animal's well-being from a psychological perspective?
3. What is meant by *promote*? The dictionary indicates that this means "to help bring into being."
4. What is *well-being*? The dictionary says "the state of being happy, healthy, or prosperous." Is the use of the word *or* significant in the application of this definition?

This whole discussion may seem like splitting hairs, but when we are dealing with language that may be pivotal in the development and interpretation of federal standards, it is necessary to pay close attention to detail. In order to adequately address the role of research results in a particular regulatory issue, we must first thoroughly understand the intent behind the legislative language. Evaluating legislative language and responding to proposed regulations often requires the precise attention to detail that our mentors tried to pound into our heads during our training.

Revised requirements for cage sizes and access to exercise space are prominently featured in the proposed standards concerning the psychological well-being of primates. Why would this aspect of animal care be the main focus of this regulatory process? Well, if we look at how this language became incorporated in the amended act, we can begin to realize the origin of the idea that bigger cages must be better.

At the NABR conference mentioned above, the senator who sponsored this language commented that he visited a laboratory housing a small number of chimps. The facility was located in a large, sparsely populated western state where space has never been in short supply. To provide the animals with more space in this facility was a fairly simple solution to the perceived problem. But the same cannot be said for many of the locations currently housing nonhuman primates.

A further question is: Does this statute really mean primates in the all-inclusive sense? If so, will the National Institutes of Health pay for office furniture, employee lounges, and vacations to more enjoyable environments? Does the Animal Welfare Act (1985) prohibit a vociferous abolitionist from diminishing my psychological well-being?

I know that this type of argument borders on the absurd, but it is important that legislative directives be clear and, where they are not clear, that the lack of clarity be understood by all those involved when the development of standards is attempted. Depending upon the interpretation of this language, the standards that we would each envision could vary significantly and thus affect our input into the regulatory process.

WHY?

Why are we facing the prospect of standards from such unclear language, which provides little if any direction? This happened because the language and even the basic concept that underlies it never received a thorough review during the hearing phase that is generally included in the legislative process. In this instance, we stand along with the somewhat green postdoctoral trainee. We did not drop the ball or make a mistake. We were not given the opportunity to do so. We are being forced to provide our input after, not before, the proposed language became law. At this point we have provided written comments based upon our scientific data. If those data indicate that the proposed standards are flawed in some respect, we will be accused of arguing against improving the well-being of the animals in our care. The fact that legitimate scientific concerns were never documented for the official record in legislative hearings will put us at a significant disadvantage from the standpoint of how the public may perceive our intentions. By not being permitted to provide input at the front end of the process, we have been put in a less than optimal position.

WHAT ARE YOU (WE) GOING TO DO ABOUT IT?

The first step in developing a planned solution to a perceived problem is to evaluate the resources available. In the situation under discussion, the resources are people and their expertise. When it comes to studying, caring for, and utilizing nonhuman primates, we have access to an extremely diverse population of experienced professionals. The key word in the last statement is *diverse*. Although this diversity adds tremendous breadth to our level of expertise, it has, in the past, often served as an obstacle in formulating any type of cohesive effort in addressing issues that affected us all. This situation must be addressed if we as a scientific community are going to have meaningful input into the regulatory process. If research results are to play an integral part in the development and final implementation of the standards, all who can contribute to this process must work as a team.

Turning the regulatory process into a public debate where emotional statements, which either misrepresent or conveniently overlook the facts, are all that the public hears will have the effect of overshadowing all the scientific data directed to the United States Department of Agriculture (USDA). Our response must be both part of the regulatory process and an answer to public concerns. Before scientists and agency staff can translate research results into workable regulations, they must translate them into a language that the public understands. Comparing the wide open spaces of the western United States to the crowded streets of Manhattan is something with which the public identifies. Discussions of conspecifics, alpha males, manipulanda and multifactorial regressions are not things with which the public can easily identify.

How must the scientific community respond to the proposed standards? We must respond truthfully and with the facts that are available to us. If we do not have the necessary information to address a point, we must clearly state that such information is not available and that it needs to be developed before such standards are fully implemented.

We must stress the diversity of the behavioral repertoire of the varied species to which the standards will apply and the need for more research in this area. In addition, we must explain the complexity of research being done and the fact that this type of research takes time. We must publicly face the issue that optimally addressing the animals' well-being in some studies could create a level of nonexperimental variability that could seriously compromise the validity of those studies. We must convey the fact that an environment that carefully controls the experimental and nonexperimental variables is not necessarily one that can optimally meet the psychological needs of nonhuman primates.

In order to address these concerns, we must stress the need to develop housing systems and management strategies that contribute to the animals' well-being in an environment that can still meet the mandated requirements for proper sanitation and the provision of adequate veterinary care. We must convey the importance of considering the cost-benefit ratio of making major expenditures that may have only minimal long-term effects on the well-being of the animals. We must convey the fact that implementation of any new standards in this area will take time in each institution and that as research data become available, it may be necessary to reevaluate the standards and make changes when such changes are warranted. We must demonstrate that we are the experts and that our expertise should be given credence by responding to our concerns and not allowing an emotional campaign to stampede the regulatory process into developing standards for which there are insignificant data available.

We must convey that we have been and will continue to be concerned with animal welfare. For those of us who have chosen to work with and study animals for our life's work, nothing is a bigger personal or professional insult than to be excluded from consideration as animal welfarists. Finally, we must convey the absolutely essential need for facts. The public must be exposed to the truth in debates that affect the regulatory process.

In summary, we must communicate the complexity of this issue to the public, to our legislative representatives, to the regulatory agencies, and to our colleagues. We must clearly demonstrate our concern for the well-being of the animals in our charge, while addressing the need to define, from a scientific perspective, what represents well-being and how it can be achieved in an environment designed to minimize nonexperimental variability.

The development and implementation of highly subjective standards may produce costs that far exceed the initial expense of new cages and enrichment devices, if the initial evaluation of the benefits derived by the animals follows, rather than proceeds implementation.

Translating research results into workable regulations is part of an overall process of educating the public, our elected representatives, and the regulatory agencies. This process should be one that is ongoing in nature and addresses the issues discussed during this conference. The scientific community must gain the public's confidence and respect as a group of individuals who also care before we can interject our facts into this emotionally charged issue.

Reference

Animal Welfare Act, 7 U.S.C. §§ 2131-2157 (1985).

CHAPTER 20

PUBLIC PERCEPTION CONCERNING RESEARCH WITH CAPTIVE PRIMATES AND OTHER ANIMALS

W. JEAN DODDS

Most scientists involved in biomedical research depend upon the use of whole animals or of animal tissues and fluids to serve as surrogates or "models" for questions directed at understanding human health and disease. Although studying naturally occurring animal diseases should ultimately benefit the animals as well as humans with analogous disorders, extrapolating from experimentally induced conditions in animals to human diseases is inherently more problematic from both scientific and ethical perspectives. Scientists have therefore evolved an increasing self-awareness and sensitivity to the need for justifying the use of animals and for reducing the degree of harm inflicted on them for the advancement of knowledge. This process has been termed "self-regulation."

The success of the self-regulatory approach to issues of animal welfare in relation to scientific needs has been exemplified over the past decade by the formation of new organizations directed toward this goal. One such organization is the Scientists' Center for Animal Welfare (SCAW).

SCAW was founded in 1979 by a group of individuals concerned with the need to forge a middle ground between the goals of scientific advancement of knowledge and of human responsibility to protect the welfare of research animal subjects. SCAW is a unique nonprofit organization of scientists and supportive members who care about animals and about how they are used for the benefit of people and other animals. This care includes special concern for all animals used in necessary but responsible research.

SCAW accepts that there is a need to use animals in certain types of scientific studies and other human endeavors. Through educational activities and scientific scholarship, it promotes responsible and caring treatment of laboratory, farm, and wild animals involved in research, and discourages undesirable and redundant research that employs animals.

Through discussion and study, SCAW helps define policy for and the accountability and responsibility attendant upon persons who use animals for any purpose. Thus, SCAW seeks the best possible reconciliation of human needs with the requirements and well-being of all animals and by its commitment to a scientific, objective approach to resolving differences provides a constructive forum for settling issues related to animal well-being. SCAW thereby occupies the reasoned middle ground on often contentious and polarized animal welfare or animal rights controversies.

The above overview reflects the policies and beliefs of SCAW, which are shared by many other scientists and concerned individuals. Collectively, have we succeeded in achieving even some of our stated goals in the last decade? Does the public appreciate that a growing middle-ground consensus is developing around these issues? The answer is clearly affirmative as attested to by the fact that we have come a long way both societally and academically in achieving these goals over the past decade (Dodds & Orlans, 1982; Niemi & Dodds, 1986).

In November 1981, for example, SCAW held the first of a series of national conferences on topics related to animal welfare. The meeting was held at the National 4H Center in Washington, DC, and the proceedings were published in 1982 (Dodds & Orlans, 1982). Thomas E. Malone, acting director of the National Institutes of Health (NIH), gave the keynote address, which pointed the way toward implementing refinement, replacement, and reduction in the care and use of laboratory animals. At that meeting, major emphasis was placed on discussions as to whether prior review of animal research protocols was appropriate or even feasible; and secondly whether institutions should consider having lay members serve on protocol review committees. Not surprisingly, both suggestions generated considerable alarm, and 89% of attendees who completed a postconference questionnaire evidenced a negative attitude concerning lay membership.

To put the current climate with respect to animal issues in perspective, it is useful to review briefly the major recommendations generated by SCAW's initial conference held in the early 1980s, when Harlyn O. Halvorson of Brandeis University emphasized that scientists must participate whenever public policy for the use of animals in biomedical research is being determined (Halvorson, 1982). Recently, John Kaplan of Stanford University reiterated this point (Kaplan, 1988). The involvement of science should not only include participation by individual scientists and scientific societies but should also address the public image of science, contributions of basic science, mechanisms whereby public policy is generated, social responsibility, and finally a cooperative approach to ensure that ethical considerations for animals will reflect the views of the various groups that contribute to the larger society.

Twelve recommendations emanated from the first SCAW conference (Dodds & Orlans, 1982). Many of these have since been endorsed or adopted, an accomplishment of which the biomedical community, public, government, and the many individuals who contributed can justly be proud. These recommendations (adapted from Dodds & Orlans, 1982, pp. 123–126) are worthy of reiterating to help focus our goals for the 1990s:

1. Scientists must work continuously toward elimination or prevention of unnecessary pain for research animals.
2. Investigators should discuss their research proposals with other responsible scientists before submitting the protocol to the animal care and use committee (ACUC) for approval.
3. ACUCs should include members with a broad representation of viewpoints (e.g., technicians and graduate students, community representatives and ethicists).
4. The public should be made more aware of why and how research is conducted on experimental animals.
5. Funding agencies should require investigators applying for grants to address animal issues specifically.
6. Public and private funding agencies should use consultants with expertise in animal issues to review selected grant proposals that present special concerns.
7. Training courses should be provided to scientists to increase their empathy concerning the animals they are using, as well as their knowledge of established animal-care policies.
8. Accreditation of institutions by the American Association for the Accreditation of Laboratory Animal Care (AAALAC) should be fostered.
9. Inspection procedures by the federal government should be improved.
10. "Codes of practice" should be established for all journals expressing editorial policy with reference to humane guidelines for care and use of experimental animals.
11. The ability of journal editors to assess animal welfare measures accurately should be enhanced by:
    a. publishing guidelines for the welfare of experimental animals in each journal;

   b. stipulating that reviewers appraise submitted papers for compliance with appropriate animal welfare standards; and
   c. using a numerical classification system for animal experimental procedures based on the degree of intervention and physical or psychological trauma to the animal.

12. More funds should be allocated by public and private sources for upgrading animal facilities.

At subsequent conferences during the ensuing 5 years, there were prolonged discussions of how ACUCs would separate scientific merit issues from those of animal welfare, care, and research design (Orlans, Simmonds, & Dodds, 1987). Only recently have committees begun to accept that it is often impossible to separate totally these issues that, by their very nature, tend to be interdependent. At some institutions merit review is performed independently at the departmental level, and a summary is provided to assist the ACUC. In other instances, merit review discussions are left primarily to external review groups such as NIH Study Sections.

A final prevailing theme throughout the growth phase of this national endeavor has been balancing the cost-benefit equation from an ethical perspective. At what point does the ethical cost or burden for the animal outweigh the anticipated benefit to society from performing the research? In balancing this delicate equation, basic common sense should prevail. Similarly, whenever there is doubt, the scale should always be tipped in favor of the least harm to the animal (Orlans et al., 1987).

The consensus recommendation from the most recent national SCAW conference (Orlans et al., 1987) included 18 points, which primarily restated and expanded upon the original recommendations published 5 years earlier. Today self-regulation and awareness are steadily improving. Scientists are more amenable and even eager to comply with the current atmosphere of protocol review and modification—especially if they can be shown that another way will achieve their scientific goals more humanely. Perhaps the most effective strides toward a more humane ethic can be made in refining protocols to reduce animal harm, along with implementation of an earlier endpoint than death to alleviate suffering.

However, lest we rest on our laurels or believe that the scientific community has adequately understood and addressed the issues, we need only observe the mounting animal advocacy and activist movement in our country (Kaplan, 1988; National Academy of Sciences, 1988). Society accords us the privilege of performing scientific investigations with animals, especially when use of public funds is involved. This privilege cannot be taken lightly, as we are accountable for our actions and attitudes.

Scientists prefer to debate issues with facts and reasoned viewpoints. Nevertheless, if we are to make credible headway in convincing the public of our commitment to the principles of good science, good sense, and good sensibility (Orlans et al., 1987), our actions need to match our words. Do we truly accept our role as stewards for the animals we study? Do we exercise respect and compassion for animals, and have we made the necessary pledge to ensure that animals in our care are free from harm and suffering? If the answer is yes, then we should stop using inappropriate language by referring to research animals as "tools, models, subhuman primates, or preparations." This objectification of laboratory animals fosters an ambivalence about their true value (Arluke, 1988). Additionally, why do we tolerate "speciesism" among scientists and others when it comes to protecting the welfare of rats and mice versus companion animals and captive primates? Rodents constitute more than 80% of the total number of animals used in biomedical research in this country, yet rats and mice still are not included as named species listed in the Animal Welfare Act, despite the fact that the Secretary of Agriculture is empowered to do so. Nevertheless, the scientific community would do well to put aside lingering negatives and excuses for lethargy or apathy on animal welfare issues. Let's have a bias for action—where there's a will there's a way!

References

Arluke, A. B. (1988). Sacrificial symbolism in animal experimentation: Object or pet? *Anthrozoos*, *2*, 97–116.

Dodds, W. J., & Orlans, F. B. (Eds.). (1982). *Scientific perspectives on animal welfare*. New York: Academic Press.

Halvorson, H. O. (1982). The responsibility of scientists in determination of public policy for the use of animals in biomedical research. In W. J. Dodds & F. B. Orlans (Eds.), *Scientific perspectives on animal welfare* (pp. 109–121). New York: Academic Press.

Kaplan, J. (1988). The use of animals in research. *Science*, *242*, 839–840.

National Academy of Sciences - National Research Council. (1988). The use of laboratory animals in biomedical and behavioral research. Washington, DC: National Academy Press.

Niemi, S. M., & Dodds, W. J. (1986). Animals in research. *National Forum*, *66*, 25–29.

Orlans, R. B., Simmonds, R. C., & Dodds, W. J. (Eds). (1987). Effective animal care and use committees. *Laboratory Animal Science*, Special Issue, 178.

CHAPTER 21

THE PSYCHOLOGICAL WELL-BEING OF CAPTIVE PRIMATES: PROTECTING THE PUBLIC INTEREST

EDWARD A. LEONARD

The mechanism for the enforcement of animal-welfare regulations in the Commonwealth of Massachusetts is an unusual one in two ways. First, the Commissioner of Public Health, who is responsible for animal welfare under state law, licenses local humane societies to inspect animal facilities housing dogs or cats, to report their findings to the state, and to take necessary action to enforce the regulations. In this way a nongovernment, public organization that is chiefly concerned with animal welfare becomes the guarantor of the public interest in the care and use of animals in research and other institutions in the state.

Second, inspectors are authorized to inspect *any and all* animals held in that facility, whether or not they are specifically named in the law. Therefore, Massachusetts inspectors *do* have the authority to inspect and regulate facilities using rodents, farm animals, birds, and other species excluded at present from the federal animal-welfare regulations.

This system of regulation, in which research institutions are monitored by animal-welfare organizations under the rules of a state agency, raises two questions: Does this system provide better assurance of the well-being of captive primates and other laboratory animals? Could this arrangement be employed successfully for monitoring federal standards? We will return to these questions after a consideration of the process of animal-welfare inspection and regulation in Massachusetts.

As a veterinarian for the Animal Rescue League of Boston, I wear many hats. I am the director of veterinary medicine for the League's clinic and shelters. I am also a special police officer for the League, licensed by the Massachusetts Department of Public Safety. In addition, I conduct inspections of research institutions, as authorized by the Commissioner of Public Health. The color of these hats I wear may change with your point of view.

LICENSING AND INSPECTING RESEARCH FACILITIES IN MASSACHUSETTS

In accordance with Massachusetts General Law, chapter 140, section 174D, the Commissioner of Public Health issues licenses to research institutions that use dogs or cats in scientific investigations, experiments, or instruction, or for the testing of drugs or medicine. The Commissioner grants licenses to fit and proper research institutions, if the issue thereof is in the public interest. The standards, buildings, routines, and activities assure the fitness of the institution.

The Commissioner makes rules and regulations needed to carry out the aims of the law, and he, or an agent chosen by him, visits and inspects the animal research and care facilities. As agents for the Commissioner, two animal-welfare organizations, the Animal Rescue League of Boston and the Massachusetts Society for the Prevention of Cruelty to Animals, conduct visits and inspections. According to the law, "animal" refers to dogs and cats primarily and to all other sentient creatures except humans. Inspections for the granting, extending, or renewal of a license involve all these animals at the facility.

As an inspector, I visit facilities without prior notice at least four times a year. I inspect the facilities, review records, and ask about practices and activities relating to animal care, handling, or veterinary treatment. Inspections are conducted most often during normal business hours. However, they may take place outside of ordinary business hours, if necessary.

Any institution I inspect provides me with the location of all animals owned, held, or in its custody. I tour the buildings and rooms that hold animals for research or testing. At the end of the inspection, I meet with the veterinarian in charge of the facility, or his or her designee, and report my findings and comments. Reported findings are those that appear to constitute faults in compliance with regulations from the Commissioner.

STANDARDS IN MASSACHUSETTS

The Massachusetts regulations for research institutions set rules for licensing. They also ensure the humane treatment, specifically of dogs and cats, but also of all other animals in general. As with other rules that guide institutions, these provide both facility operating standards and animal health and husbandry standards.

The building, water supply, power, storage, waste disposal, and even the cleanliness of caretakers must be adequate. Indoor facilities need heat, ventilation, lights, and waterproof surfaces. Outdoor structures give shelter from sun, rain, snow, and cold weather. Both indoor and outdoor facilities need proper drainage.

Enclosures must provide for the animals' health and comfort with adequate space. They must allow each animal to turn about, stand, sit, lie, and stretch freely in a comfortable normal position. Amended federal regulations on larger cage size will replace any smaller requirements that the state may use currently.

The animal health and husbandry standards describe the feeding of healthy, tasty food in sufficient quantity. Safe drinking water in clean, handy containers must be provided. Housing must be clean, prevent contamination, and reduce disease hazards and odors. The process of sanitizing primary enclosures, cages, rooms, and hard-surfaced pens or runs must physically clean, remove, and wipe out unhealthy agents to the most practical extent possible.

Housekeeping procedures should protect the animals from injury and disease. Pest control must be effective, and records must be kept. Enough trained employees must be available to keep up the prescribed level of husbandry practices. Finally, plans for good veterinary care and attention must be set up and maintained.

UNIQUE ASPECTS OF STATE LAW

These portions of the state research animal regulations are akin to those in the federal Laboratory Animal Welfare Act. However, the Department of Public Health provides additional requirements similar to some of those of the Public Health Service and other agencies. For example, in Massachusetts, animals likely to suffer more than fleeting pain *must* have the use of proper drugs in contrast to the federal statute that makes the use of pain-relieving drugs discretionary by the attending veterinarian. Any process likely to cause more discomfort than that of anesthetization calls for anesthesia. Postsurgical care must include repeated observations of the animal's condition, and any action needed to assure comfort and ease postsurgical pain must be taken.

If anesthesia would defeat the purpose of the experiment, and there is no less painful procedure that might be used, then the regulations require prior approval of the procedure in writing. Approved projects that do not meet these standards for pain control must abide by the National Institutes of Health (NIH) "Guide for the Care and Use of Laboratory Animals" (National Research Council, 1985). All unnecessary pain and injury must be avoided, and the experiment must be ended when unnecessary pain may result. Each licensee must treat all research animals in a manner consistent with the intent of these rules.

A MODEL FOR RESEARCH REGULATION?

Now let me return to the two questions we considered at the outset: Does the system used in Massachusetts provide better assurance of the well-being of captive primates and other laboratory animals? Could this arrangement be successful for federal standards?

To answer the first question, consider an organization marked by kindness, mercy, and concern for animals. Allow this animal-welfare organization to check carefully the facilities, methods, and animal care of a research institution. This seems to many people to be a good way to satisfy public concerns. It can remove doubts about conflicts of interest between public and private funding agencies, regulators, and the scientific community.

The second question is more difficult. Scientists may have reason to fear the present Massachusetts arrangement. Could such an arrangement be the beginning of a mixture of state and local regulations enforced by local humane society and animal rights groups? An important point here is whether the humane society views research as a valid use of animals. Do these organizations want the continuation of proper animal research, but under the best possible setting? Or do they seek to halt all animal use through more strict regulation and inquiry?

I cannot say that the Animal Rescue League of Boston supports or approves of animal research. The League supports and approves of animal welfare. The League is also a law-and-order animal-welfare society. We have police officers to enforce the cruelty laws in the courts. Our legislative aides support or oppose proposed laws and amendments as needed to promote animal welfare. Because the scientific research community supports and approves some of the same animal welfare aims, then we have common goals to pursue.

If the scientist and the animal activist agree on the welfare goals for animals, would the government agencies accept civilian inspectors? This type of arrangement would take much time and discussion to satisfy all those interested.

I am sure each inspector would have to complete a training course approved by the agency before being authorized to conduct inspections. The training program in Massachusetts was the result of combined efforts. State, federal, research, academic, and animal welfare organizations all contributed. As a result of this effort, I think our inspectors better understand two very important principles. First, the standards set the rules for the inspection. Second, and perhaps more important, the goal of the inspection is to ensure the humane treatment of the animals Because of this, Massachusetts has a successful program that does help assure laboratory animal well-being.

Reference

National Research Council. (1985). *Guide for the care and use of laboratory animals* (NIH Pub. No. 85–23). Washington, DC: U.S. Department of Health and Human Services.

CHAPTER 22

THE PUBLIC'S PERCEPTION OF PRIMATES IN RESEARCH

STEVE CARROLL

PROMOTING ANIMAL RESEARCH

The Incurably Ill for Animal Research (iiFAR) is a national organization with local chapters throughout the country representing (a) those suffering with illnesses and disorders for which there currently are no cures; (b) those whose lives have been saved by recent medical developments such as organ transplants and cancer treatments made possible by animal research; and (c) healthy people who wish to see medicine progress so that their children and grandchildren might be spared from acquired immune deficiency syndrome (AIDS), heart disease, cancer, and other deadly ailments. In short, we represent the beneficiaries of research, the people for whom biomedical research is being conducted.

In their efforts to promote the continuation of animal research, our members interact directly with the general public on a day-to-day basis. This interaction puts iiFAR in a unique position to observe the public's reactions to, and misperceptions about, the use of animals in medical research. One of the areas where these misperceptions are the greatest is the use of primates.

It is important for those of us working to promote the continuation of animal research to understand how the public views animals used in medical research. It is certain that the public relations people working with the animal rights organizations understand this, and they use it to their advantage every day.

For example, anyone subject to animal-activist propaganda has heard that all animals used in research are "pets" of one type or another. Recently on the

Sally Jessy Raphael show, the host began the program by stating that cats, dogs, monkeys, and rabbits are the animals used in research. Cleveland Amory talked about the "thousands of kittens" shot by the military and all the "puppies" that are force-fed oven cleaner. And, because everyone knows that chimpanzees are 3 feet tall, wear clothes, and eat with silverware, People for the Ethical Treatment of Animals (PETA) recently had such a young animal, wearing a PETA T-Shirt, as the "host" of a congressional luncheon.

PERCEPTIONS OF ANIMALS

The animal activists focus on dogs, cats, and primates, because they realize the public has a deeply ingrained affection for these animals, while also having a deeply ingrained indifference (or even dislike) for the other common research animals. After all, very few people are willing to open the doors of their homes to rats and mice. By understanding how the public views research animals, particularly primates, we can better target our educational programs and, even more importantly, avoid reinforcing the public's misperceptions.

To understand why the public views primates the way it does, we need to look at the way these animals have been presented to us over the years. We have always been fascinated with primates. After all, in many ways they look and act like humans. In recent years, our schools have taught that we are descended from nonhuman primates, and that living primates are in some ways our "cousins." This tendency to overidentify with nonhuman primates is enhanced by the fact that, unlike all other common research animals, primates are not native to this country. The average person has had no natural exposure to them. We only see them in artificial settings, where many times they are being financially exploited for our amusement.

Who cannot remember the excitement when the circus came to town? As a child I watched the chimps and monkeys, all dressed in brightly colored clothes, playing catch with the clowns and riding the horses just like real cowboys (complete with cowboy hats and six-shooters). I especially enjoyed watching little "Bingo" outsmart Lou Costello on the Abbot and Costello television show. There were many similar television shows and movies during the 1950s, and there are some even today. Even when my mother would make me watch educational programs about nature on public television, I learned from them that female gorillas care for their young "just like human mothers."

Our zoological parks today do an excellent job of displaying animals in surroundings that are as close as possible to the animals' natural habitat. But when I was a child, I remember going to the small local zoo, where they had one mean-looking gorilla (like King Kong) in a cage and also a compound full of chimpanzees. All the chimps were about 3 feet tall, and their compound was almost an exact duplicate of my grade-school playground. The chimps even had "monkey bars" to play on.

Many people working on promoting the continuation of animals in research have compared the primate issue with that of the pound issue. In many ways, there are similarities. Dogs, cats, and primates all evoke strong emotions in people—and they are favorite subjects for the animal activists. However, in some ways the primate issue is very different and presents special problems that need to be considered.

Most people have direct contact with cats and dogs, but not with primates. Whether they live in rural or urban settings, most people in this country have a rather balanced exposure to companion animals—cats and dogs. Anybody living in a city has had dogs get into the trash. Everybody has been kept up at night by the mating cries of cats. It is not uncommon to see on television or read in the paper about a dog's biting a child and the child's having to go through a long and painful series of rabies treatments. So people are somewhat exposed to the negative side of companion animals.

But most people don't get that balanced exposure to primates. They only view primates in a very positive way—they're cute, cuddly, and lovable This perception is distorted. The fact that a chimp, or even a "cute little monkey," could be a dangerous animal is something that most people find very hard to accept. Many find it inconceivable that a chimpanzee is capable of causing physical harm to a human being.

From a public relations standpoint, there are a lot of problems in dealing with this attitude, just as there are many problems in trying to define and measure the psychological well-being of primates, as many papers in this volume show. I wish there were a simple answer, but there isn't. However, if we realize the position from which the general public views primates, we can better target our educational efforts so as not to reinforce these misperceptions.

SELF-HELP FOR RESEARCHERS

Our organization talks with the public daily about the use of animals in research, and part of our work concerns discussing the use of primates in research. As scientists, you too should become involved in similar programs—in grassroots, public education programs, per se. When you do, there are a few pointers that I would like to offer from our perspective, based on our experience and the feedback we have received.

First, we all need to be very careful about how we draw comparisons between human and nonhuman primates. We must not reinforce the stereotyped humanization of nonhuman primates and other animals. That is not to say that there are never instances when discussing human and nonhuman similarities is the proper thing to do. However, this is something that we need to think about and do carefully. For example, we might avoid using the classic "people" words when discussing any animals, such as referring to their offspring as "kids." Even when we use an appropriate scientific term with a precise meaning, such as "juveniles," we may be unwittingly reinforcing this humanization of the animal kingdom.

Second, we need to be careful about the photographs that we use to illustrate the lives of primates. For example, it is common for photographs of primates to contain commercially available and highly recognizable toys that are marketed for human children. Depicting such toys may help to illustrate important aspects of primate behavior, but it also reinforces stereotypes and misperceptions. It is another example of projecting and reinforcing an image that is inaccurate.

Third, we need to identify clearly the numbers of primates used in research. Many people believe that dogs, cats, rabbits, and primates are the only animals used in medical research, and it is important that such people get a proper perspective on which animals are used, how they are used, and how many are used.

Fourth, it is very important to stress that we are *all* concerned about the primates' well-being, that we do respect their well-being, and that well-being is always considered when choosing which animals to use and how they will be used. We should always be prepared to respond to the questions about why this research is being done with primates or why it cannot be done with other, non-animal methods or with species other than primates.

Whenever possible, the point should be made that research helps not only humans, but has immediate and tangible benefits to nonhuman primates as well. In fact, we should point out those cases in which the benefits go beyond the laboratory and are of wider importance, perhaps including species still in the wild. Many people are concerned about dwindling populations of primates, and much of what goes on in labs, zoos, and universities can benefit those animals still in the wild.

Fifth, it is wisest to present adult animals when dealing with the media and the general public, rather than young animals. In using young animals, we reinforce the myth that the 3-foot chimpanzee is a charming human child.

Sixth, we should always be ready to deal with the importation questions and the issue of whether the use of primates in research is helping to drive the wild populations into extinction.

CONCLUSION

The use of primates in research is an issue that evokes strong emotions in the general public. It is one that presents peculiar problems from a public relations standpoint. There is a lot of misinformation; there are a lot of misperceptions that need to be addressed. All professional organizations concerned with research and concerned with the availability of primates for biomedical and behavioral research need to work within their own frameworks and work together in conferences like this one to overcome these misperceptions. The next step is to bring together people involved in communicating directly with the public on animal research issues, so that we can have the opportunity to share our experiences.

CHAPTER 23

ANIMAL RESEARCH: OUR OBLIGATION TO EDUCATE

FREDERICK A. KING

SCIENCE AND MODERATION

As a consequence of opposition from anti-animal-research organizations, the continued existence of primate research will depend upon a major, sustained public information campaign conducted with the full participation and cooperation of all the life sciences. Some may argue that the ongoing changes in regulations following the 1985 amendments to the Animal Welfare Act are unnecessary, because animal research in the main has been conducted humanely and judiciously. It has been said that the amendments and the currently proposed regulations, especially those concerned with the "psychological well-being of primates," are designed primarily to pacify the animal activists, and once these regulations are in place, all will be well for animal research.

There is convincing evidence that this will not be the case. Based upon their own statements and past performance, we may be assured that zealous leaders and the extremist organizations in the antivivisectionist and animal rights movement will not be satisfied unless there is a total ban on biomedical and behavioral research with primates, as well as with other laboratory animals.

Support from Grant RR-00165 from the Division of Research Resources, National Institutes of Health to the Yerkes Regional Primate Research Center is acknowledged. The Yerkes Center is fully accredited by the American Association for Accreditation of Laboratory Animal Care.

The extreme position of the animal-rights activists is not new. More than 9 years' experience dealing directly with research animal issues and anti-animal-research organizations has shown me the futility of attempts to reach accords or accommodations between the values and practices of the animal-research community and those of the antivivisectionists and animal-rightists. As Larry Horton, Associate Vice President for Public Affairs at Stanford University, has pointed out: "Moderates from the animal welfare movement have not been much in evidence. . . . Those who profess moderation often support the same legislation as out-and-out antivivisectionists" (Horton, 1987, p. 13). So we may ask the question—where have all the moderates gone?

The answer to this question is that the scientists, in their responsible concern for the welfare of their animal subjects and their desire to minimize pain and provide care, occupy that middle ground. But in the absence of genuine extremists who would defend willful mistreatment of animals and fight against any regulation whatsoever, scientists have been deliberately mislabeled by the antivivisectionists as insensitive and licentious—that is, at the extreme of the animal-welfare spectrum. However, the reality is, as Horton has said, "No one on the biomedical side represents an extreme position" (Horton, 1987, p. 14).

PUBLIC OPINION AND EDUCATION

The public remains insufficiently informed of the benefits that animal research has brought to human and animal health. It is the responsibility of the scientific community to inform the public of precisely how animal research has benefitted both human and animal health, and of the ways in which animal care and experimentation are regulated by federal, state, and local authorities. Although public opinion polls (Research Strategies Corp., 1985) indicate that a significant majority of the people in this country are not opposed to animal research and indeed are cognizant of its need, it is essential for us to recognize that many people still have doubts about its necessity and about the way it is conducted.

Public information about animal research comes mainly from those groups who are vehemently opposed to it. This fact represents an educational tragedy. A prominent public relations consultant once told me, "People are down on what they are not up on." People are down on, or at least questioning, the necessity of research with primates and other laboratory animals and the way it is conducted, because, in a large measure, either they are not aware of the essential role animals have played in research leading to human benefits, or they do not realize the strict federal, state, and local regulations and inspections with which animal-research facilities and programs must comply (Research Strategies Corp., 1985).

The anti-animal-research community is doing a much better job in influencing the public than we are. With their multimillion dollar treasuries (Anderson

& Spear, 1988; Anderson & Van Atta, 1988; Hendee, Loeb, Schwarz, & Smith, 1989; Mobilization for Animals, 1983), these groups bombard the public with costly full-page advertisements in major newspapers and magazines, frequent and constant mailings of newsletters and brochures, television and radio announcements, videotapes, movies, protest rallies, concerts—and other means of publicity. Further, the animal-rights community has its own national magazines. *The Animals' Agenda* is professionally produced and distributed free to high-school libraries, members of Congress, and the news media ("Outward Bound," 1987; "Sustaining a Newsmagazine," 1986). An even glossier and more expensive publication, *The Animals' Voice*, has recently emerged.

In 1978, only a dozen or so registered animal rights activists and antivivisectionists lobbied in Washington. Today, many more are active in pressing for restrictive legislation on animal research. The pressure the anti-animal-research groups place on Congress is immense. No wonder science is often on the losing side when the animal-rightists have many millions of dollars, a mass of full-time employees, and the enduring dedication to attack us relentlessly.

Even some legislators have succumbed to propaganda by the antiresearch community. One elected representative to the U.S. Congress recently wrote in a letter to his California constituents: "It is my belief that the continued use of animals in experimentation is an example of lazy methodology and an inability to break from traditional information-gathering techniques" (Dellums, 1988). This letter, displaying a remarkable lack of information and understanding, was reprinted in the newsletter of In Defense of Animals, whose president wrote in the same issue:

> We are in a war. It is a no-holds-barred war, a dirty war where administrators and researchers alike will use any tactic to win. Misinformation, ridicule, dishonesty, charges of fraud will become weapons that the animal exploiters will use with increasing frequency as our efforts become stronger and more effective (Katz, 1988, p. 13).

From this falsity it should be abundantly clear that scientists are untruthfully accused of the very tactics employed routinely by their opponents.

If animal-activist groups today were genuinely concerned with improving conditions in the laboratory, would they have done everything in their power to block and delay new and improved animal facilities at Stanford University, the University of California at Berkeley, and the University of California at San Francisco? This was done at a cost of millions of wasted dollars and years of deliberate and malicious obstruction of improvements that would benefit the animals themselves as well as the research (Chui, 1988; Holden, 1989; Kennedy, 1987). If their goal was the improvement of animal care and health, would the Animal Liberation Front (recognized as a terrorist organization by the FBI and the Attorney General's Office of California), which is represented to the public by the People For the Ethical Treatment of Animals (PETA) (Brown, 1983; Griswold, 1985; Harvard University, 1982) and other extremist groups (Bishop, 1988), have burned to the ground, at a cost of millions of dollars, a new clinic and veterinary research unit at the University of California at Davis and vandalized buildings at the University of Arizona? The public must be told that it is the *scientists* who occupy the moderate middle ground and the leaders of the animal rights and antivivisectionist movements who are clearly the destructive extremists.

In their zeal to persuade the public that animal research is useless and cruel, these groups make minimal use of the facts and maximal use of distortion and emotional appeal. They also take advantage of the lack of public understanding about the nature and process of science (McDonald, 1989). President Donald Kennedy of Stanford University has referred to "the disappointing level of scientific literacy displayed by the voters—and by their elected representatives" (Chui, 1988, p. 1232).

The failure of our educational system to foster an understanding of the role science has played in the development of civilization and in freeing us from the darkness of medieval ignorance contributes to the public's lack of awareness of the health benefits that science has provided to humans and animals. Failure to understand the process of science has resulted in a vulnerability of the public to the arguments of the anti-animal-research movement—a movement which at heart is fundamentally antiscientific, anti-intellectual, and antihumanistic. If successful, this movement will weaken and very possibly destroy the very fabric of the biological, medical, and behavioral sciences. Each of you who works in the area of animal science is under attack. It is you, your work, and the present and future victims of illness who will suffer if scientists do not work together to inform the public, our legislators, and administrators of the value of programs related to animal research.

Science today would be in a much stronger position with the public if individual researchers and their professional and scientific organizations had come down from their ivied towers 20 or more years ago to speak clearly about their work and its benefits. We should all have been explaining long ago what it is we do, how we do it, and why. The time has come, or rather is long past, for us and our institutions to explain clearly and practically (a) how discoveries through basic research with animals have laid the groundwork for remarkable advances in both human and animal health and (b) how and why primates and other laboratory animals are used to test the safety and effectiveness of drugs and to develop vaccines and new treatments.

For too many years, scientists, physicians, and their institutions have frequently downplayed their animal studies, apparently believing that the public would not understand or care. This silence was a grievous error for which we are paying dearly today. The public does care, as shown by the upsurge in the animal-rights movement and its increased public acceptance. You may be assured that the public is not only interested in how research animals are treated but also in the implications of our science as well. It is up to us to take the time to explain, for science is not immune from public concern, scrutiny, and accountability. The people of our nation, who as taxpayers ultimately subsidize most of our research, have a right to be informed of what goes on in science, of how we treat the animals in our charge, and of the rewards to society that research brings.

PUBLIC EDUCATION THROUGH THE UNIVERSITY

We need much better, broader, and more comprehensive educational programs about science and animal research. Scientific organizations are poor in fiscal resources compared to the animal rights groups, who have many millions of dollars available. Hence, we need more intense and sustained participation by individual scientists, animal caregivers, technicians, veterinarians, and clinicians. Public relations and public policy experts say that the success of an educational campaign depends on adequate grassroots efforts. Our universities are in an ideal position to conduct and coordinate such grassroots efforts, because they still enjoy immense prestige and credibility and exist for the purpose of education. Let them then educate on behalf of the laboratory animal research that they advocate and conduct.

Universities have also the unique ability to reach a variety of extremely important audiences, including their own students, faculty, and staff. We must press and stimulate our universities, with help and guidance from national scientific organizations, to provide the necessary resources, courses, lecture series, discussion groups, and the like, to inform students about the methods, history, regulation, and benefits of animal research. Humane regulations for the care and use of animals in research that do not hinder science should be a major concern of every biological, medical, and behavioral scientist and of every student.

A comprehensive educational program about animal research is needed, but with special emphasis on primates and pound animals, because these are especially targeted by the animal-rightists. I do not propose an educational campaign solely for primate research. Primate studies are not conducted in isolation from research with other species, and the public's lack of understanding of animal research is certainly not restricted to primate research.

Such a campaign should stress concrete examples of the benefits of primate research in which all of us share. These examples can provide potent and convincing arguments if they are used well and frequently. Recently, after a national animal-rights organization paid for full-page ads in *The New York Times* and *The Washington Post* about the so-called horrors of research with primates and other animals (The New England Anti-Vivisectionist Society, 1988), it was suggested that the scientific community sponsor ads stating, "Thirty years ago, research with monkeys helped produce the polio vaccine. Today, studies with primates may provide a vaccine and drugs to combat AIDS,'' or ''Primates constitute three percent of the laboratory animals needed in research. But what an important three percent!'' Very few of such ads have reached the public, unfortunately, because we still do not have a sufficiently well-funded or effective national campaign to explain research to the public, despite the efforts of such organizations as the National Association for Biomedical Research (NABR).

Although national scientific and professional organizations still need to become more involved in explaining primate research to the public, local university efforts are crucial in influencing public opinion, and it is at this grassroots level that scientists can be the most active. We desperately need the participation of individual scientists. I do not suggest that individual researchers go off entirely on their own to conduct educational campaigns. As individuals, we do not have the time, money, or expertise to accomplish that sort of education. Our participation should be as part of university and institutional programs.

Accurate information and instruction about the care and use of animals in research should be a part of our undergraduate and graduate life science courses and especially a part of the curricula of the colleges of education that are preparing those who will teach science to our children in the public schools. Teachers of science in elementary and high schools are extraordinarily important targets for a university's educational efforts, because the anti-animal-research organizations are already strongly and effectively carrying their message to them (NABR, 1988).

Whenever possible, institutions using animals in research should provide briefings to local grade school and high school biology teachers and students about animal research, its benefits, and how it is regulated. The Yerkes Primate Research Center conducts such presentations to school science clubs and local science teachers. The Oregon Primate Research Center does even more: It sponsors a summer program that brings local school teachers to the center for varying periods of time to observe, participate, and learn about animal research.

Universities and colleges have many other audiences that they can easily reach, on and off campus. These include their own employees and alumni as well as patients and their family members at university-affiliated hospitals and clinics. Virtually all our universities publish newsletters or magazines, hold tours and briefings on alumni day, and produce brochures for patient waiting rooms. These and other established university public relations programs can and should be used to educate people about animal research.

A great educational goal could be achieved if, in each university, the deans and vice presidents for research and medicine and directors of research institutions would include in all news releases about important advances in human health some mention of the animal contributions that made the human benefits possible. For example, the prowess and skills of organ transplant surgeons are impressive and deserving of headline recognition. But a half-century and more of research in cellular biology, chemistry, and immunology, most of it with animals, has played an equal role in making transplants possible. The public should be told of the basic-science animal studies that were essential for these achievements.

At the local level, civic organizations often turn to their universities as sources of speakers for their meetings. Universities can and should offer to them speakers to address topics of animal research, care, and concerns. In doing so, universities and scientists need to be more assertive in describing the opponents of science.

HUMANIZING THE ISSUES

The public deserves to know that the animal-rights movement is not, at heart, concerned with improving animal-research facilities and welfare, but with stopping research. Behind apparent pleas for humane treatment are frequent lies, distortions, and a radical philosophy exemplified by the statement of Ingrid Newkirk of the People for the Ethical Treatment of Animals that ‘‘a rat is a pig is a dog is a boy’’ (McCabe, August 1986) and her odious comparison of animal research and the killing of chickens for food to the Nazi holocaust (Brown, 1983). Newkirk and others who have so frequently used the Nazi reference have an inverted and totally uninformed historical perspective that is highly offensive to humankind. As Larry Horton (1988) well noted:

> The Nazi reference is dead wrong for two reasons: (1) it ignores the antivivisectionist strain in Nazi philosophy and policy; and (2) it ignores the actual conclusions drawn from Nazi brutality and embodied in the postwar Nuremberg Code.
>
> In the first year the Nazis came to power, they took decisive action on antivivisection. An early Nazi propaganda piece extolling Hitler's virtues includes this accolade: ‘‘Your Fuhrer is the strongest opponent of any form of animal torment; especially vivisection, which is the ‘scientific’ torture of animals, and a heinous product of the Jewish materialistic school of medicine, about which he states: ‘in the national-socialist state, this situation will end very soon.’ ’’
>
> In a 1933 press release by the Reich Press Agency of the NSDAP, the importance of the Nazi policy was proclaimed as follows: ‘‘among all civilized states, Germany is the first country to end the shame of vivisection. The New Germany not only frees people from the curse of materialism, egotism and cultural bolshevism, but also gives rights to the tortured, tormented and, up until now, completely unprotected animals. The animal friends and opponents of vivisection in all countries will hail this act of the National Socialist government of the New Germany with a joy! What Reich Chancellor Adolf Hitler and Prime Minister Goering did and will do for the protection of animals stands as a guideline to the leaders of all civilized states.’’

> We know all too well the truth about Nazi experiments on humans in death camps. The postwar examination of what happened there resulted in the promulgation of the Nuremberg Code, which explicitly states that human experiments should be based on the results of animal experimentation. For those with a fresh understanding of Nazi policies, animal research was not an atrocity; it was a moral obligation.
>
> Yet what has been called the Auschwitz analogy—comparing research with animals to the holocaust—is given common currency, sometimes by Members of Congress, sometimes by others. Here is an interesting twist added by Ingrid Newkirk, referred to in *The Washington Post* as the Joan of Arc of Animal Rights: "Six million people died in concentration camps, but six billion broiler chickens will die this year in slaughterhouses." I leave it to you to evaluate the moral content of that statement. (pp. 41–44)

To compare the Holocaust with animal research is false and deceiving, because all of Western civilization militates against such a comparison of human and animal life, and because the Nazis placed animal life above human life.

Finally, scientists and universities need to humanize animal-rights issues by involving patients as spokespersons, because they are the most immediate actual and potential benefactors of animal research. We must ally ourselves with organizations such as that remarkably effective and dedicated group known as the Incurably Ill For Animal Research (iiFAR) (Carroll, 1988); and, with all compassion for animals, take the emotionalism out of the arguments on research with dogs, cats, and primates and put it where it belongs—in concern for the human beings who desperately need and can benefit from animal research.

Obviously some institutions do their share, but far too many continue with their heads in the sand with the futile hope that the problem will disappear of its own accord (Holden, 1989). It will not. If we wish to see medical, biological, and behavioral studies with animals survive and prosper, it is not only a necessity, but our obligation, to lobby our own universities and institutions to become actively involved in what they do best—education—in this case, for animal research. Let us convince our institutions to persuade the public to share our vision and understand how continued progress in medicine and mental health depends upon animal research.

References

Animal Welfare Act, 7 U.S.C. §§ 2131-2157 (1985).

Anderson, J., & Spear, J. (1988, September 8). "Excessive" pay at humane society. *The Washington Post*, p. A15.

Anderson, J., & Van Atta, D. (1988, September 7). Dubious deals in the humane society. *The Washington Post*, p. A17.

Bishop, K. (1988, January 19). Rise in militancy seen by animals advocates. *The New York Times*, p. 46.

Brown, C. (1983, November 13). She's a portrait of zealotry in plastic shoes. *The Washington Post*, p. B1.
Carroll, S. (1988, January 9). Primates in research. *iiFARsighted Report*, *2* pp. 1–8.
Chui, G. (1988). Activists beset UC, Stanford labs. *Science*, *239*, 1232.
Dellums. (1988). In defense of animals. *Perspective*, *3*, 3.
Griswold, C., Jr. (1985). Ingrid Newkirk. *City Paper*, *5*, 37–48.
Harvard University, Office of Government and Community Affairs. (1982). *The animal rights movement in the United States: Its composition, funding sources, goals, strategies and potential impact on research*. Cambridge, MA: Harvard University Press.
Hendee, W. R., Loeb, J. M., Schwarz, M. R., & Smith, S. J. (1989). Use of animals in biomedical research: The challenge and response. *AMA White Paper*, p. 8.
Holden, C. (1989). Universities fight animal activists. *Science*, *243*, 17–19.
Horton, L. (1987). *Putting the politics of research with animals in perspective: Remarks to an NIH workshop on the implementation of public health service policy on humane care and use of laboratory animals*. Unpublished manuscript.
Horton, L. (1988). A look at the politics of research with animals: Regaining lost perspective. *The Physiologist*, *31*, 41–44.
Katz, E. M. (1988). In defense of animals. *Perspective*, *3*, 13.
Kennedy, D. (1987, October 29). The anti-scientific method. *The Wall Street Journal*, p. 26.
McCabe, K. (1986, August). Who will live, who will die? *The Washingtonian*, pp. 112-157.
McDonald, K. A. (1989, January 25). One in 18 Americans is found "scientifically literate." *The Chronicle of Higher Education*, A7.
Mobilization for Animals. (1983). Animal welfare organizations: A first annual financial report (p. 3). Columbus, OH.
National Association for Biomedical Research. (1988). Annual report (p. 14). Washington, DC.
New England Anti-Vivisection Society. (1988, July 17). Reading the truth about animal testing will take five minutes and a very strong stomach. *The New York Times*. (p. 5).
Outward bound. (1987). *The Animals' Agenda*, *7*, 2.
Research Strategies Corp. (1985). *Members of the American public comment on the use of animals in medical research and testing: A study conducted for the Foundation for Biomedical Research*. Princeton, NJ: Author.
Sustaining a newsmagazine for animals. (1986). *The Animals' Agenda*, *6*, 2.

APPENDIX A

THE WELFARE OF NONHUMAN PRIMATES IN RESEARCH SETTINGS: A BIBLIOGRAPHY

MELINDA A. NOVAK, KARLA H. DREWSEN, AND SUSAN A. BECKLEY

INTRODUCTION

The following is an extensive bibliography (over 650 articles) on the welfare of nonhuman primates in research settings. The bibliography is divided into a number of different sections including ethics, laws, and psychological well-being. However, most of the articles are contained in the last two sections of the document, which are concerned with those manipulations that affect the well-being of animals in laboratory settings. The section on physical environment includes references on such topics as cage design, cage size, and environmental enrichment (e.g., novel objects, foraging devices, mirrors, etc.), and the section on social environment is concerned with rearing experiences, group formation, group composition, and reintroductions of hand-reared infants to conspecifics.

CONTENTS

ETHICS

A sampling of articles and books concerned with the ethics of animal use in research. Topics include animal sentience and its implication for research protocols, animal rights, researcher responsibilities, anthropomorphism and speciesism, the morality and utility of animal research.

Adams, P. M. (1981). The scientist's concern for animal welfare. *Psychopharmacology Bulletin, 17*, 91–93.

American Medical Association. (1988). *White Paper: Use of animals in biomedical research, the challenge and response*. Washington, DC: American Medical Association.

Archer, J. (1986). Ethical issues in psychobiological research on animals. *Bulletin of the British Psychological Society, 39*, 361–364.

Archibald, J., Ditchfield, J., & Rowsell, H. C., (Eds.). (1985). *The contribution of laboratory animal science to the welfare of man and animals: Past, present and future*. New York: Gustav Fischer Verlag.

Banks, E. M. (1982). Behavioral research to answer questions about animal welfare. *Journal of Animal Science, 54*, 434–446.

Bannister, D. (1981). The fallacy of animal experimentation in psychology. In D. Sperlinger (Ed.), *Animals in research: New perspectives in animal experimentation* (pp. 307–318). New York: John Wiley & Sons.

Bowd, A. D. (1980). Ethical reservations about psychological research with animals. *Psychological Record, 30*, 201–210.

Boyce, J. R. (1989). Use of animals in research: Can we find a middle ground? *Journal of the Veterinary Medical Association, 194*, 24–25.

Britten, S. (1977). Animal research: Whose responsibility? *World Medical Journal, 24*, 51–53.

Burghardt, G. M. (1985). Animal awareness: Current perceptions and historical perspective. *American Psychologist, 40*, 905–919.

Caplan, A. L. (1978). Rights language and the ethical treatment of animals. In L. McCullough & J. P. Morris (Eds.), *Implications of history and ethics to medicine—veterinary and human* (pp. 126–135). College Station: Texas A & M University Press.

Caplan, A. L. (1983). Beastly conduct: Ethical issues in animal experimentation. In J. A. Sechzer (Ed.), *The role of animals in biomedical research. Annals of the New York Academy of Science, 406,* 159–169.

Clark, S. R. L. (1982). *The nature of the beast: Are animals moral?* New York: Oxford University Press.

Clark, S. R. L. (1984). *The moral status of animals*. New York: Oxford University Press.

Cohen, C. (1986). The case for the use of animals in biomedical research. *New England Journal of Medicine, 315,* 865–870.

Coile, D. C., & Miller, N. E. (1984). How radical animal activists try to mislead humane people. *Laboratory Primate Newsletter, 23,* 11–13.

Cooper, T., & Stucki, J. C. (1983). Commentary on "animal research"—for and against: A philosophical, social and historical perspective. *Perspectives in Biology and Medicine, 27,* 18–21.

Dawkins, M. (1980). *Animal suffering*. New York: Chapman Hall.

Dewey, J. (1926). The ethics of animal experimentation. *Atlantic Monthly, 138,* 343–346.

Dewsbury, D. A. (1990). Early interactions between animal psychologists and animal activists and the founding of the APA Committee on Precautions in Animal Experimentation. *American Psychologist, 45,* 315–327.

Diamond, C. (1981). Experimenting on animals: A problem in ethics. In D. Sperlinger (Ed.), *Animals in research: New perspectives in animal experimentation* (pp. 337–362). New York: John Wiley & Sons.

Dodds, W. J., & Orlans, F. B. (Eds.). (1982). *Scientific perspectives on animal welfare*. New York: Academic Press.

Dresser, R. (1988). Standards for animal research: Looking at the middle. *The Journal of Medicine and Philosophy, 13,* 123–143.

Drewett, R., & Kani, W. (1981). Animal experimentation in the behavioral sciences. In D. Sperlinger (Ed.), *Animals in research: New perspectives on animal experimentation* (pp. 175–201). New York: John Wiley & Sons.

Driscoll, J. W., & Bateson, P. (1988). Animals in behavioural research. *Animal Behaviour, 36,* 1569–1574.

Erichsen, S. (1984). How do I justify my use of animals for experimental purposes? A researcher's view. *Canadian Association of Laboratory Animal Science Newsletter, 16,* 58–59.

Eudey, A. A. (1981). Ethical concerns in primate use and husbandry. *International Journal for the Study of Animal Problems, 2,* 96–102.

Feeney, D. M. (1987). Human rights and animal welfare. *American Psychologist, 42,* 593–599.

Fox, M. W. (1978). Humane ethics in veterinary education. *Journal of Veterinary Medical Education, 5,* 98–101.

Fox, M. W. (1981). The question of animal rights. *Veterinary Record, 109,* 37–39.

Fox, M. W. (1981). Experimental psychology, animal rights, welfare and ethics. *Psychopharmacology Bulletin, 17,* 80–84.

Fox, M. W. (1983). Humane ethics and animal rights. *International Journal for the Study of Animal Problems, 4,* 286–289.

Fraser, A. (1975). Ethology and ethics. *Applied Animal Ethology, 1,* 211–212.

Frey, R. G. (1983). Vivisection, morals and medicine. *Journal of Medical Ethics, 9,* 94–97.

Gallup, G. G., & Beckstead, J. W. (1988). Attitudes toward animal research. *American Psychologist, 43,* 474–475.

Gallup, G. G., & Suarez, S. D. (1980). On the use of animals in psychological research. *Psychological Record, 30,* 211–218.

Gallup, G. G., & Suarez, S. D. (1985). Alternatives to the use of animals in psychological research. *American Psychologist, 40,* 1104–1111.

Gay, W. I. (1986). *The health benefits of animal research.* Washington, DC: Foundation for Biomedical Research.

Gilmore, A. (1985). The use of animals in research. *Canadian Medical Association Journal, 132,* 564–568, 570–571.

Goodall, J. (1987). A plea for the chimpanzees. *American Scientist, 75,* 574–577.

Harken, D. E., & Harken, A. H. (1989). Anti-antivivisection, have we waited too long? *Archives of Surgery, 124,* 1366–1367.

Held, J. R. (1980). Responsibilities of the investigator in the utilization of laboratory animals. *Development of Biological Standards, 45,* 213–219.

Held, J. R. (1983). Appropriate animal models. In J. A. Sechzer (Ed.), *The role of animals in biomedical research. Annals of the New York Academy of Science, 406,* 13–19.

Held, J. R., & Milouchine, V. (1981). Use and protection of laboratory animals. *Animal Regulatory Studies, 3,* 273–279.

Herzog, H. A. (1988). The moral status of mice. *American Psychologist, 43,* 473–474.

Hodos, W. (1983). Animal welfare considerations in neuroscience research. In J. A. Sechzer (Ed.), *The role of animals in biomedical research. Annals of the New York Academy of Science, 406,* 119–127.

Hoff, C. (1980). Immoral and moral uses of animals. *New England Journal of Medicine, 302,* 115–118.

Horton, L. The enduring animal issue. *Journal of the National Cancer Institute, 81,* 736–743.

Iglehart, J. K. (1985). The use of animals in research. *New England Journal of Medicine, 313,* 395–400.

Jacobs, F. S. (1989). Animal rights: A biological contradiction. *Journal of the American Veterinary Medical Association, 195,* 10–11.

Johnson, D. (1990). Animal rights and human lives: Time for scientists to right the balance. *Psychological Science, 1,* 213–214.

Kanigel, R. (1987, January/February). Specimen No. 1913. *The Sciences,* pp. 30–37.

Kaplan, J. (1988). The use of animals in research. *Science, 242,* 839–840.

Keehn, J. D. (1977). In defense of experiments with animals. *Bulletin of the British Psychological Society, 30,* 404–405.

King, F. A. (1984). Animals in research: The case for experimentation. *Psychology Today, 18,* 56–58.

King, F. A., & Yarborough, C. J. (1985). Medical and behavioral benefits from primate research. *The Physiologist, 28,* 75–87.

Kronfeld, D. S., & Parr, C. P. (1987). Ecologic and symbiotic approaches to animal welfare, animal rights, and human responsibility. *Journal of the American Veterinary Medical Association, 191,* 660–664.

Lane-Petter, W. (1976). The ethics of animal experimentation. *Journal of Medical Ethics, 2,* 118–126.

Lansdell, H. (1988). Laboratory animals need only humane treatment: Animal "rights" may debase human rights. *International Journal of Neuroscience, 42,* 169–178.

Leader, R. W., & Stark, D. (1987). The importance of animals in biomedical research. *Perspectives in Biology and Medicine, 30,* 470–485.

Levine, M. E. (1977). Animal rights evaluated. *The Humanist, 37,* 12–15.

Locke, E. A. (1989). Animal rights advocates: Are they threatening human rights? *Lab Animal, 18,* 19.

Magel, C. R. (1981). *A bibliography of animal rights and related matters*. Washington, DC: University Press of America.

March, B. E. (1984). Bioethical problems: Animal welfare and animal rights. *Bioscience, 34*, 615–620.

Markell, D. L. (1981). The case for revising our laws on animal experimentation. *International Journal for the Study of Animal Problems, 2*, 87–95.

McCabe, K. (1986, August). Who will live, who will die? *The Washingtonian*, pp. 112–118, 153–157.

Midgley, M. (1981). Why knowledge matters. In D. Sperlinger (Ed.), *Animals in research: New perspectives on animal experimentation* (pp. 319–336). New York: John Wiley & Sons.

Miller, N. E. (1983). Understanding the use of animals in behavioral research: Some critical issues. In J. A. Sechzer (Ed.), *The role of animals in biomedical research. Annals of the New York Academy of Science, 406*, 113–118.

Miller, N. E. (1984). Value and ethics of research on animals. *Laboratory Primate Newsletter, 23*, 1–10.

Miller, N. E. (1985). The value of behavioral research on animals. *American Psychologist, 40*, 423–440.

Moss, T. H. (1984). The modern politics of animal use. *Bioscience, 34*, 621–626.

Newman, A. (1989). Research versus animal rights: Is there a middle ground? *American Scientist, 77*, 135–137.

Nicoll, C. S., & Russell, S. M. (1990). Editorial: Analysis of animal rights literature reveals the underlying motives of the movement: Ammunition for counter offensive by scientists. *Endocrinology, 127*, 985–989.

Paton, W. (1984). *Man and mouse: Animals in medical research*. New York: Oxford University Press.

Rajecki, D. W. (Ed.). (1983). *Comparing behavior: Studying man, studying animals*. Hillsdale, NJ: Erlbaum.

Regan, T. (1983). *The case for animal rights*. Berkeley, CA: University of California Press.

Regan, T., & Singer, P. (Eds.). (1976). *Animal rights and human obligations*. Englewood Cliffs, NJ: Prentice-Hall.

Ritvo, H. (1984). Plus ca change: Antivivisection then and now. *Bioscience, 34*, 626–633.

Rollin, B. E. (1981). *Animal rights and human morality*. Buffalo, NY: Prometheus Books.

Rollin, B. E. (1985). The moral status of research animals in psychology. *American Psychologist, 40*, 920–926.

Rosner, F. (1985). Is animal experimentation being threatened by animal rights groups? *Journal of the American Medical Association, 254*, 1942–1943.

Ross, M. W. (1981). The ethics of experiments on higher animals. *Social Science Medicine, 5*, 51–60.

Rowan, A. N. (1984). *Of mice, models, and men: A critical evaluation of animal research*. Albany, NY: SUNY Press.

Rowan, A. N., & Rollin, B. E. (1983). Animal research—for and against: A philosophical, social, and historical perspective. *Perspectives in Biology and Medicine, 27*, 1–17.

Rowsell, H. C. (1977). The ethics of biomedical experimentation. In *The future of animals, cells, models, and systems in research, development, education and testing*. Washington, DC: National Academy of Sciences.

Russell, J. C., & Secord, D. C. (1985). Holy dogs and the laboratory: Some Canadian experiences with animal research. *Perspectives in Biology and Medicine, 28*, 374–381.

Sechzer, J. A. (1983). The ethical dilemma of some classical animal experiments. In J. A. Sechzer (Ed.). *The role of animals in biomedical research. Annals of the New York Academy of Science, 406*, 5–12.

Shapiro, H. M. (1986). Animal rights and biomedical research: No place for complacency. *Anesthesiology, 64*, 142–146.

Short, R. V. (1986). Primate ethics. In K. Benirschke (Ed.), *Primates the road to self-sustaining populations* (pp. 1–13). New York: Springer-Verlag.

Shuster, S. (1977, November). Why we need animals in research. *World of Medicine*, pp. 19–37.

Sieber, J. E. (1986). Students' and scientists' attitudes on animal research. *The American Biology Teacher, 48*, 85–91.

Singer, P. (1975). *Animal liberation: A new ethics for our treatment of animals.* New York: New York Review.

Smith, S. J., & Hendee, W. R. (1988). Animals in research. *Journal of the American Medical Association, 259*, 2007–2008.

Snyder, M. (1989, January 18). We have a moral obligation to continue conducting research involving animals. *The Chronicle of Higher Education.*

Soave, O. (1981). Animal rights, animal welfare, and the law. *California Veterinarian, 35*, 27–29.

Sperlinger, D. (1981). Natural relations—contemporary views of the relationship between humans and other animals. In D. Sperlinger (Ed.), *Animals in research: New perspectives in animal experimentation* (pp. 79–104). New York: John Wiley & Sons.

Taylor, P. W. (1986). *Respect for life: A theory of environmental ethics.* Princeton, NJ: Princeton University Press.

Uvarov, O. (1984). Research with animals: Requirement, responsibility, welfare. *Laboratory Animals, 19*, 51–75.

Vaughn, C. (1988). Animal research: Ten years under siege. *Bioscience, 38*, 10–13.

Verderber, R. F. (1984). Depolarization of humane animal care groups. *Lab Animal, 13*, 47–51.

Visscher, M. B. (1975). *Ethical constraints and imperatives in medical research.* Springfield, IL: Charles C. Thomas & Co.

Wall, P. D. (Ed.). (1982). Editor's note to readers and authors on the ethics of animal experiments. *Pain, 12*, 199.

White, R. J. (1971). Antivivisection: The reluctant hydra. *American Scholar, 40*, 503–507.

Zola, J. C., Sechzer, J. A., Sieber, J. E., & Griffin, A. (1984). Animal experimentation: Issues for the 1980s. *Science, Technology, and Human Values, 9*, 40–50.

LAWS AND REGULATIONS

Legal Issues:

Barany, E. (1983). The Swedish system of ethical committees in the laboratory animal field. *Conquest, 172*, 10–15.

Black, H. S., & Doepel, F. M. (1985). Legislation, policy, attitudes impacting ARF management. *Lab Animal, 14*, 40–43.

Blackman, D. E. (1981). Regulating psychological experimentation with animals in the United Kingdom. *Psychopharmacology Bulletin, 17*, 84–88.

Broad, W. J. (1980). Legislating an end to animals in the lab. *Science, 208*, 575–576.

Chambers, K. T., & Hines, C. (1983). Recent developments concerning the use of animals in medical research. *Journal of Legal Medicine, 4*, 109–129.

Curran, W. J. (1984). Biomedical laboratories and criminal liability for cruelty to animals. *Lab Animal, 13*, 18–21.

Favre, D. S., & Loving, M. (1983). *Animal law*. New York: Oxford University Press.

Fitzgerald, T. A. (1980). Implementation of regulations for the protection of man and animals used in biomedical research at a United States school of medicine. *Development of Biological Standards, 45*, 203–212.

Flowers, F. H. (1983). Research animal care in Canada: Its control and regulation. In J. A. Sechzer (Ed.), *The role of animals in biomedical research. Annals of the New York Academy of Science, 406*, 144–149.

Goldsmith, E. I., & Moor-Jankowski, J. (1980). International movement for humane principles in animal experimentation. *Journal of Medical Primatology, 9*, 1–4.

Griffin, A., & Sechzer, J. A. (1983). Mandatory versus voluntary regulation of biomedical research. In J. A. Sechzer (Ed.), *The role of animals in biomedical reserarch, Annals of the New York Academy Science, 406*, 187–200.

Hampson, J. E., & Silcock, S. R. (1985). Animal protection and toxicity testing regulations. *Food Chemistry and Toxicology, 23*, 183–186.

Johnson, D. K., & Morin, M. L. (1983). U.S. laws, regulations and policies important to managers of nonhuman primate colonies. *Journal of Medical Primatology, 12*, 223–238.

Lagoni, N., Fiebelkorn, J., & Wormuth, H. J. (1983). Protection of animals and animal experimentation: A survey of scientific experts. *International Journal for the Study of Animal Problems, 4*, 225–228.

Markell, D. L. (1981). The case for revising our laws on animal experimentation. *International Journal for the Study of Animal Problems, 2*, 87–95.

Mattingly, S. F. (1984). Should humane care standards be reviewed? *Lab Animal, 13*, 25–31.

McCarthy, C. R. (1984). What the Public Health Service is doing to improve humane care. *Lab Animal, 13*, 21–26.

Melyers, N. M. (1983). Government regulation of nonhuman primate facilities. *Journal of Medical Primatology, 12*, 169–183.

Obrink, K. J. (1983). Are regulations to protect experimental animals adequate? *Progress in Clinical and Biological Research, 128*, 159–166.

Orlans, F. B. (1984). What institutional animal research committees can do to improve humane care. *Lab Animal, 13*, 24–29.

Overcast, T. D., & Sales, B. D. (1985). Regulation of animal experimentation. *Journal of the American Medical Association, 254*, 1944–1949.

Parson, R. M. (1983). National and international regulations governing transportation and the supply of primate animals. *Journal of Medical Primatology, 12*, 262–266.

Paton, W. D. (1983). Animal experiment: British and European legislation and practice. In J. A. Sechzer (Ed.), *The role of animals in biomedical research, Annals of the New York Academy of Science, 406*, 201–214.

Purchase, I. F. (1984). The future of animals in research and teaching. The European scene. *Fundamentals of Applied Toxicology, 4*, 519–530.

Rowan, A. N. (1981). Scientific issues and regulation of primate use. *International Journal for the Study of Animal Problems, 2*, 37–43.

Rowsell, H. C. (1980). The voluntary control program of the Canadian council on animal care. *Journal of Medical Primatology, 9*, 5–8.

Schwindaman, D. F. (1983). The Animal Welfare Act as applied to primate animal laboratories. *Journal of Medical Primatology, 12*, 250–255.

Stewart, W. C. (1984). Legal standards for humane care: The Animal Welfare Act. *Lab Animal, 13*, 33–41.

Vallier, G. (1980). European concepts on the use of laboratory animals in relationship with animal welfare problems. *Development of Biological Standards, 45*, 189–195.

Guidelines:

Animal Behaviour Society. (1986). Animal care guidelines. *Animal Behaviour, 34*, 315–318.

Association of American Medical Colleges; Association of American Universities. (1986). Recommendations for governance and management of institutional animal resources. *Lab Animal, 15*, 34–38.

American Physiological Society. (1984). Guiding principles in the care and use of animals. *Physiologist, 27*, 141.

American Psychological Association. (1985). *Guidelines for ethical conduct in the care and use of animals.* Washington, DC: American Psychological Association.

Burke, R. E., Edinger, H. M., Foreman, D. S., Hand, P. J., Hodos, W., Irwin, L. N., Kelly, D. D., King, F. A., Miles, F. A., & Morrison, A. R. (1984). Guidelines for the use of animals in neuroscience research. *Neuroscience Newsletter, 15*, 1–4.

Canadian Council on Animal Care. (1980). *Guide to the care and use of experimental animals*, (Vol. 1). Ottawa, ON: Canadian Council on Animal Care.

Council of Europe. (1985). *European convention for the protection of vertebrate animals used for experimental or other scientific purposes.* Strasbourg, France: Council of Europe.

Guidelines for the use of animals in research. (1986). *Quarterly Journal of Experimental Psychology, 38B*, 111–116.

Larson, J. (1988). Services of the Animal Welfare Information Center. *Journal of the American College of Toxicology, 7*, 262–265.

National Research Council. (1985). Guide for the care and use of laboratory Animals. (NIH Pub. No. 85–23). Washington, DC: U.S. Department of Health and Human Services.

New York Academy of Sciences. (1988). *Interdisciplinary principles and guidelines for the use of animals in research, testing, and education.* New York: New York Academy of Sciences.

Orlans, F. B. (Ed.). (1988). *Field research guidelines*. Bethesda, MD: Scientists Center for Animal Welfare.

Porta, M., Nyberg, C. R., & Boast, C. (1989). Laboratory animal welfare: Keeping up. *Lab Animal, 18*, 21–27.

Sanford, J., Ewbank, R., Molony, V., Tavernor, W. D., & Uvarov, D. (1986). Guidelines for the recognition and assessment of pain in animals. *Veterinary Record, 118*, 334–338.

Smith, A. W., Houpt, K. A., Kitchell, R. L., Kohn, D. F., McDonald, L. E., Passaglia, M., Thurmon, J. C., & Ames, E. R. (1986). Report of the AVMA panel on euthanasia. *Journal of the American Veterinary Medical Association, 188*, 252–268.

Working Committee for the Biological Characterization of Laboratory Animals/GV-SOLAS. (1985). Guidelines for the specification of animals and husbandry methods when reporting the results of animal experiments. *Laboratory Animals, 19*, 106–108.

U.S. Laws:

Animal Welfare Act, Pub. L. No. 91–579, §2, 84 Stat. 1560 [Codified as amended beginning at 7 U. S.C. §2131 (1970)].

Animal Welfare Act, Pub. L. No. 94–279, 90 Stat. 417 [Codified as amended at 7 U. S. C. §§2131–2156 (1976)].

Animal Welfare Act, 7 U. S. C. §§2131–2157 (1985).

Laboratory Animal Welfare Act, Pub. L. No. 89–544, §1, 80 Stat. 350 [Codified as amended beginning at 7 U. S. C. §2131 (1966)].

U.S. Dept. of Agriculture, Animal and Plant Health Inspection Service, Proposed rule (9 C.F.R. 1–3) [Docket No. 88–103]. Federal Register, 54 (49), 10822–10954; March 15, 1989.

U.S. Dept. of Agriculture, Animal and Plant Health Inspection Service, Definition of Terms and Regulations, Final Rule (9 C.F.R. 1;2) [Docket No. 89–130; 89–131]. Federal Register, 54 (168), 36122–36163; Aug. 31, 1989.

U.S. Dept. of Agriculture, Animal and Plant Health Inspection Service, Animal Welfare; Standards; Proposed Rule (9 C.F.R., Part 3) [Docket No. 90–040]. Federal Register, 55, (158), 33448–33531; Aug. 15, 1990.

See Also:

Genoways, H. H., & Choate, J. R. (1976). Federal regulations pertaining to collection, import, export, and transport of scientific specimens of mammals. *Journal of Mammalogy, 57*(Suppl.), 1–9.

Johnson, D. K., & Morin, M. L. (1983). U.S. laws, regulations, and policies important to managers of nonhuman primate colonies. *Journal of Medical Primatology, 12*, 223–238.

PSYCHOLOGICAL WELL-BEING OF CAPTIVE PRIMATES

A selection of articles that specifically address the issue of psychological well-being or ill-being and its measurement in captive primates. This section is concerned with well-being as a dependent variable. Subsequent sections on the physical and social environment are concerned with those environmental factors (independent variables) that may influence psychological well-being.

Bayne, K. (1989). Resolving issues of psychological wellbeing and management of laboratory nonhuman primates. In E. F. Segal (Ed.), *Housing, care, and psychological wellbeing of captive and laboratory primates* (pp. 27–39). Park Ridge, NJ: Noyes.

Bernstein, I. S. (1989). Breeding colonies and psychological well-being. *American Journal of Primatology, Suppl. 1*, 31–36.

Bramblett, C. A. (1989). Mental wellbeing in anthropoids. In E. F. Segal (Ed.), *Housing, care, and psychological wellbeing of captive and laboratory primates* (pp. 1–11). Park Ridge, NJ: Noyes.

Coe, C. L., & Scheffler, J. (1989). Utility of immune measures for evaluating psychological well-being in nonhuman primates. *Zoo Biology, Suppl. 1*, 89–99.

Curtis, S. E. (1985). What constitutes animal well-being? In G. A. Moberg (Ed.), *Animal stress* (pp. 1–14). Bethesda, MD: American Physiological Society.

Dantzer, R. (1989). Assessment of psychological well-being in animals: Lessons from farm animal studies. *American Journal of Primatology, Suppl. 1*, 5–8.

Duckelow, W. R., & Duckelow, K. B. (1989). Reproductive and endocrinological measures of stress and nonstress in nonhuman primates. *American Journal of Primatology, Suppl. 1*, 17–24.

Duncan, I. J. H. (1978). The interpretation of preference tests in animal behaviour. *Applied Animal Ethology, 4*, 197–200.

Fouts, R. S., Abshire, M. L., Bodamer, M., & Fouts, D. H. (1989). Signs of enrichment: Toward the psychological well-being of chimpanzees. In E. F. Segal (Ed.), *Housing, care, and psychological wellbeing of captive and laboratory primates* (pp. 376–388). Park Ridge, NJ: Noyes.

Hearn, J. P., & Dixon, A. F. (1984). Assessment of comfort and well-being in New World primates. In *Standards in laboratory animal management* (Part 2, pp. 206–216). Potters Bar, UK: Universities Federation for Animal Welfare.

Kaplan, J. R. (1986). Psychological stress and behavior in nonhuman primates. In G. Mitchell & J. Erwin (Eds.), *Comparative primate biology*, Vol 2, Part A, pp. 455–492). New York: Alan R. Liss.

Mason, W. A. (1989). Primatology and primate well-being. *American Journal of Primatology, Suppl. 1*, 1–5.

Mench, J. A., & Krulisch, L. (Eds.). (1990). *Well-being of nonhuman primates in research*. Bethesda, MD: Scientists Center for Animal Welfare.

Moberg, G. A. (1985). Biological response to stress: Key to assessment of animal well-being? In G. A. Moberg (Ed.), *Animal stress* (pp. 27–49). Bethesda, MD: American Physiological Society.

Morton, D. B., & Griffiths, P. H. M. (1985). Guidelines on the recognition of pain, distress, and discomfort in experimental animals: An hypothesis for assessment. *Veterinary Record, 116*, 431–436.

Nadler, R. D. (1989). The psychological well-being of captive gorillas. In E. F. Segal (Ed.), *Housing, care, and psychological wellbeing of captive and laboratory primates* (pp. 416–420). Park Ridge, NJ: Noyes.

Novak, M. A. (1989). Psychological well-being: Applications to social groups of nonhuman primates. In J. W. Driscoll (Ed.), *Animal care and use in behavioral research: Regulations, issues and applications* (pp. 81–90). Beltsville, MD: USDA, National Agricultural Library.

Novak, M. A., & Drewsen, K. H. (1989). Enriching the lives of captive primates: Issues and problems. In E. F. Segal (Ed.), *Housing, care, and psychological wellbeing of captive and laboratory primates* (pp. 161–182). Park Ridge, NJ: Noyes.

Novak, M. A., & Suomi, S. J. (1988). Psychological well-being of primates in captivity. *American Psychologist, 43*, 765–773.

Snowdon, C. T., & Savage, A. (1989). Psychological well-being of captive primates: General considerations and examples from callitrichids. In E. F. Segal (Ed.), *Housing, care, and psychological wellbeing of captive and laboratory primates* (pp. 75–88). Park Ridge NJ: Noyes.

Thomas, R. K., & Lorden, R. B. (1989). What is psychological wellbeing? Can we know if primates have it? In E. F. Segal (Ed.), *Housing, care, and psychological wellbeing of captive and laboratory primates* (pp. 1–11). Park Ridge, NJ: Noyes.

Vandenbergh, J. G. (1989). Issues related to "psychological well-being" in nonhuman primates. *American Journal of Primatology, Suppl. 1*, 9–16.

Van Rooijen, J. (1984). Possibilities and limitations of choice tests in relation to animal well-being. *Proceedings of the International Congress on Applied Ethology of Farm Animals* 353–357.

Wemelsfelder, F. (1984). Animal boredom: Is a scientific study of the subjective experiences of animals possible? In M. W. Fox & L. D. Mickley (Eds.), *Advances in animal welfare science* (pp. 115–154). The Netherlands: Martinus Nijhoff.

Woolverton, W. L., Ator, N. A., Beardsley, P. M., & Carroll, M. E. (1989). Effects of environmental conditions on the psychological well-being of primates: A review of the literature. *Life Sciences, 44*, 901–917.

PHYSICAL ENVIRONMENT: ENRICHMENT OF ANIMAL HOUSING

An examination of the methods and procedures used to make the physicial environment of laboratory primates more complex and/or more natural (see also PSYCHOLOGICAL WELL-BEING).

General References:

Benirschke, K. (Ed.). (1986). *Primates: The road to self-sustaining populations.* New York: Springer-Verlag.

Bennett, C. L., & Davis, R. T. (1989). Long-term animal studies. In E. F. Segal (Ed.), *Housing, care, and psychological wellbeing of captive and laboratory primates* (pp. 213–234). Park Ridge, NJ: Noyes.

Bramblett, C. A. (1989). Enrichment options for guenons in the laboratory. *American Journal of Primatology, Suppl. 1*, 59–63.

Chamove, A. S., & Anderson, J. R. (1989). Examining environmental enrichment. In E. F. Segal (Ed.), *Housing, care, and psychological wellbeing of captive and laboratory primates* (pp. 183–202). Park Ridge, NJ: Noyes.

Erwin, J., & Deni, R. (1979). Strangers in a strange land: Abnormal behaviors or abnormal environments? In J. Erwin, T. L. Maple, & G. Mitchell (Eds.). *Captivity and behavior* (pp. 1–28). New York: Van Nostrand Reinhold.

Fajzi, K., Reinhardt, V., & Smith, M. D. (1989). A review of environmental enrichment strategies for singly caged nonhuman primates. *Lab Animal, 18*, 23–35.

Forthman Quick, D. L. (1984). An integrative approach to environmental engineering in zoos. *Zoo Biology, 3*, 65–77.

Gilbert, S. G., & Wrenshall, E. (1989). Environmental enrichment for monkeys used in behavioral toxicology studies. In E. F. Segal (Ed.), *Housing, care, and psychological wellbeing of captive and laboratory primates* (pp. 244–254). Park Ridge, NJ: Noyes.

Hediger, H. (1964). *Wild animals in captivity*. New York: Dover Publications.

Hutchins, M., Hancocks, D., & Crockett, C. (1978). Naturalistic solutions to the behavioral problems of captive animals. *American Association of Zoological Parks and Aquariums Conference Proceedings*, 108–113.

Line, S. W. (1987). Environmental enrichment for laboratory primates. *Journal of the American Veterinary Medical Association, 190*, 854–859.

Maple, T. L., & Finlay, T. W. (1986). Evaluating the environments of captive nonhuman primates. In K. Benirschke (Ed.), *Primates: The road to self-sustaining populations* (pp. 479–488) New York: Springer-Verlag.

Markowitz, H. (1982). *Behavioral enrichment in zoo animals*. New York: Van Nostrand Reinhold.

Markowitz, H., & Line, S. (1989). Primate research models and environmental enrichment. In E. F. Segal (Ed.), *Housing, care, and psychological wellbeing of captive and laboratory primates* (pp. 203–212). Park Ridge, NJ: Noyes

Markowitz, H., & Spinelli, J. (1986). Environmental engineering for primates. In K. Benirschke (Ed.), *Primates: The road to self-sustaining populations* (pp. 489–498). New York: Springer-Verlag.

McGrew, W. C. (1981). Social and cognitive capabilities of nonhuman primates: Lessons from the wild to captivity. *International Journal for the Study of Animal Problems, 2*, 138–149.

Miller-Schroeder, P., & Paterson, J. D. (1989). Environmental influences on reproduction and maternal behavior in captive gorillas: Results of a survey. In E. F. Segal (Ed.), *Housing, care, and psychological wellbeing of captive and laboratory primates* (pp. 389–415). Park Ridge, NJ: Noyes.

O'Neill, P. (1988). Developing effective social and environmental enrichment strategies for macaques in captive groups. *Lab Animal, 17*, 23–36.

O'Neill, P. (1989). A room with a view for captive primates: Issues, goals, related research, and strategies. In E. F. Segal (Ed.), *Housing, care, and psychological wellbeing of captive and laboratory primates* (pp. 135–160). Park Ridge, NJ: Noyes.

Poole, T. B. (1988). Behaviour, housing, and welfare of non-human primates. In A. C. Beynen, & H. A. Sollevold (Eds.), *New developments in biosciences: Their implications for laboratory animal science* (pp. 231–237). Dortrecht, The Netherlands: Martinus Nijhoff.

Prince, A. M., Goodall, J., Brotman, B., Dienske, H., Schellekens, H., & Eichberg, J. W. (1989). Appropriate conditions for maintenance of chimpanzees in studies with blood-borne viruses: An epidemiologic and psychosocial perspective. *Journal of Medical Primatology, 18*, 27–42.

Roberts, J. A. (1989). Environmental enrichment, providing psychological wellbeing for people and primates. *American Journal of Primatology, Suppl. 1*, 25–30.

Rumbaugh, D. M., Washburn, D., & Savage-Rumbaugh, E. S. (1989). On the care of captive chimpanzees: methods of enrichment. In E. F. Segal (Ed.), *Housing, care, and psychological wellbeing of captive and laboratory primates* (pp. 357–375). Park Ridge, NJ: Noyes.

Segal, E. F. (Ed.). (1989). *Housing, care, and psychological wellbeing of captive and laboratory primates*. Park Ridge, NJ: Noyes.

Snowdon, C. T. (1989). The criteria for successful propagation of endangered primates. *Zoo Biology, Suppl. 1*, 149–161.

Spinelli, J. S., & Markowitz, H. (1985). Prevention of cage-associated distress. *Lab Animal, 14*, 19–21.

Turkkan, J. S., Ator, N. A., Brady, J. V., & Craven, K. A. (1989). Beyond chronic catheterization in laboratory primates. In E. F. Segal (Ed.), *Housing, care, and psychological wellbeing of captive and laboratory primates* (pp. 305–322). Park Ridge, NJ: Noyes.

Van Hoof, J. A. R. A. M. (1986). Behavior requirements for self-sustaining populations—some theoretical considerations and a closer look at social behavior. In K. Benirschke (Ed.), *Primates: The road to self-sustaining populations* (pp. 307–319). New York: Springer-Verlag.

Cage Design:

Baldwin, J. D. (1968). The social behavior of adult male squirrel monkeys (*Saimiri sciureus*) in a seminatural environment. *Folia Primatologica, 9*, 281–314.

Blackmore, W. M. (1989). Solution to psychological enhancement of the environment for the nonhuman primate. In E. F. Segal (Ed.), *Housing, care, and psychological wellbeing of captive and laboratory primates* (pp. 235–243). Park Ridge, NJ: Noyes.

Chamove, A. S., Anderson, J. R., Morgan-Jones, S. C., & Jones, S. P. (1982). Deep woodchip litter: Hygiene, feeding, and behavioural enhancement in eight primate species. *International Journal for the Study of Animal Problems, 3*, 30–38.

Coe, J. C. (1989). Naturalizing habitats for captive primates. *Zoo Biology, Suppl. 1*, 117–125.

Coe, J. C., & Maple, T. (1984). Approaching Eden: A behavioral basis for great ape exhibits. *American Association of Zoological Parks and Aquariums Conference Proceedings*, 117–128.

Dahl, J. F. (1982). The feasibility of improving the captive environments of the Pongidae. *American Journal of Primatology, Suppl. 1*, 77–85.

Dahl, J. F. (1989). An inexpensive, climate-controlled enclosure for gibbons utilizing appropriate technology. In E. F. Segal (Ed.), *Housing, care, and psychological wellbeing of captive and laboratory primates* (pp. 323–334). Park Ridge, NJ: Noyes.

Else, J. G., Tarara, R., Suleman, M. A., & Eley, R. M. (1986). Enclosure design and reproductive success of baboons used for reproductive research in Kenya. *Laboratory Animal Science, 36*, 168–172.

Epple, G. (1970). Maintenance, breeding and development of marmoset monkeys (Callithricidae) in captivity. *Folia Primatologica, 12*, 56–76.

Erwin, J., Anderson, B., Erwin, N., Lewis, L., & Flynn, D. (1976). Aggression in captive pigtail monkey groups: Effects of provision of cover. *Perceptual Motor Skills, 42*, 319–324.

Goosen, C. (1988). Developing housing facilities for rhesus monkeys: Prevention of abnormal behavior. In A. C. Beynen, & H. A. Sollevold (Eds.), *New developments in biosciences: Their implications for laboratory animal science* (pp. 67–70). Dordrecht, The Netherlands: Martinus Nijhoff.

King, J. E., & Norwood, V. R. (1989). Free-environment rooms as alternative housing for squirrel monkeys. In E. F. Segal (Ed.), *Housing, care, and psychological wellbeing of captive and laboratory primates* (pp. 102–114). Park Ridge, NJ: Noyes.

Maclean, J., Phippard, A., Garner, M., Duggin, G., Horvath, J., & Tiller, D. (1987). Group housing of hamadryas baboons: A new cage design based on field studies of social organization. *Laboratory Animal Science, 37*, 89–93.

Mallinson, J. J. C. (1975). The design of two marmoset complexes at the Jersey Zoological Park. *Annual Report of the Jersey Wildlife Preservation Trust, 12*, 21–26.

Mallinson, J. J. C. (1977). Maintenance of marmosets and tamarins at Jersey Zoological Park with special reference to the design of the new marmoset complex. In D. G. Kleiman (Ed.), *The biology and conservation of the Callithricidae*. Washington, DC: Smithsonian Institution Press.

Mallinson, J. J. C. (1980). The concept behind and design of the new gorilla environment at the Jersey Wildlife Preservation Trust. *Dodo: Journal of the Jersey Wildlife Preservation Trust, 17*, 79–85.

Mallinson, J. J. C. (1982). Cage furnishings and environments for primates with special reference to marmosets and anthropoid apes. *Zoo Federation News, 23*, (pull-out Suppl. 12): 1–8.

Maple, T. L., & Stine, W. W. (1982). Environmental variables and great ape husbandry. *American Journal of Primatology, Suppl. 1*, 67–76.

Markowitz, H. (1973). Behavioral engineering in the zoo. *Ankus, 5*, 6–11.

Markowitz, H. (1975). Analysis and control of behavior in the zoo. *Research in zoos and aquariums* (pp. 77–90). National Academy of Sciences: Washington, DC.

Markowitz, H. (1978). Engineering environments for behavioral opportunities in the zoo. *Behavior Analyst*. Spring, 34–47.

Markowitz, H., Schmidt, M. J., & Moody, A. (1978). Behavioural engineering and health in the zoo. *International Zoo Yearbook, 18*, 190–194.

McKenzie, S. M., Chamove, A. S., & Feistner, A. T. C. (1986). Floor coverings and hanging screens alter arboreal monkey behavior. *Zoo Biology, 5*, 339–348.

Pereira, M. E., Macedonia, J. M., Haring, D. M., & Simons, E. L. (1989). Maintenance of primates in captivity for research: The need for naturalistic environments. In E. F. Segal (Ed.), *Housing, care, and psychological wellbeing of captive and laboratory primates* (pp. 40–60). Park Ridge, NJ: Noyes.

Riddle, K. E., Keeling, M. E., Alford, P. L., & Beck, T. F. (1982). Chimpanzee holding, rehabilitation, and breeding: Facilities design and colony management. *Laboratory Animal Science, 32*, 525–533.

Rumbaugh, D. M. (1988). Cage design attenuates display patterns of male chimpanzees. *Zoo Biology, 7*, 177–180.

Salzen, E. A. (1989). A closed colony of squirrel monkeys for laboratory studies. In E. F. Segal (Ed.), *Housing, care, and psychological wellbeing of captive and laboratory primates* (pp. 115–128). Park Rige, NJ: Noyes.

Segal, E. (1987). Caging of laboratory monkeys: Alternatives to isolation housing in small cages. *Humane Innovations and Alternatives in Animal Experimentation: A Notebook, 1*, 6.

Snowdon, C. T., Savage, A., & McConnell, P. B. (1985). A breeding colony of cotton-top tamarins (*Saguinus oedipus*). *Laboratory Animal Science, 35*, 477–480.

Stuart, M. D. (1981). Primate locomotion patterns and captive caging. *American Association of Zoological Parks and Aquariums Conference Proceedings*, 13–18.

Tollman, S. G., & Lucas, J. W. (1979). Design and establishment of a primate troop enclosure. *South African Journal of Psychology, 9*, 138–144.

Williams, L. E., Abee, C. R., Barnes, S. R., & Ricker, R. B. (1988). Cage design and configuration for an arboreal species of primate. *Laboratory Animal Science, 38*, 289–291.

Wood-Gush, D. G. M., & Beiharz, R. G. (1983). The enrichment of a bare environment for animals in confined conditions. *Applied Animal Ethology, 10*, 209–217.

Change in Environment or Comparison of Two Environments:

Bannerjee, B. N., & Woodward, G. (1970). A comparison of outdoor and indoor housing of rhesus monkeys. *Laboratory Animal Care, 10*, 80–84.

Baulu, J., & Redmond, D. E. (1980). Social and nonsocial behaviours of sex- and age-matched enclosed and free-ranging rhesus monkeys (*Macaca mulatta*). *Folia Primatologica, 34*, 239–258.

Bowen, R. A. (1980). The behavior of three hand-reared lowland gorillas, *Gorilla g. gorilla*, with emphasis on the response to a change in accomodation. *Dodo: Jersey Preservation Trust, 17*, 63–79.

Bryant, C. E., Rupniak, N. M., & Iverson, S. D. (1988). Effects of different environmental enrichment devices on cage stereotypies and autoaggression in captive Cynomolgous monkeys. *Journal of Medical Primatology, 17*, 257–269.

Chamove, A. S., & Rohrhuber, B. (1989). Moving callitrichid monkeys from cages to outside areas. *Zoo Biology, 8*, 151–164.

Clarke, A. S., Juno, C. M., & Maple, T. L. (1982). Behavioral effects of a change in the physical environment: A pilot study of captive chimpanzees. *Zoo Biology, 1*, 371–380.

Clarke, M. R., Koritnik, D. R., Martin, L. N., & Baskin, G. B. (1989). Cage enrichment, physiology, and behavior in nursery-reared rhesus monkeys. *American Journal of Primatology, Suppl. 1*, 53–57.

Fragaszy, D. M. (1979). Titi and squirrel monkeys in a novel environment. In J. Erwin, T. Maple, & G. Mitchell (Eds.), *Captivity and behavior* (pp. 172–216). New York: Van Nostrand Reinhold.

Glatson, A. R., Geilvoet-Soeteman, E., Hora-Pecek, E., & van Hoof, J. A. R. A. M. (1984). The influence of the zoo environment on social behavior of groups of cotton-topped tamarins, *Saguinus oedipus oedipus*. *Zoo Biology, 3*, 241–253.

Goerke, B., Fleming, L., & Creel, M. (1987). Behavioral changes of a juvenile gorilla after a transfer to a more naturalistic environment. *Zoo Biology, 6*, 282–295.

Herndon, J. G., Ruiz de Elvira, M. C., Turner, J. J., & Collins, D. C. (1985). Resumption of seasonal breeding patterns in male and female rhesus monkeys transferred from an indoor to an outdoor environment. *Biology of Reproduction, 32*, 733–744.

Keith-Lucas, T., & Vick, L. G. (1983). From outdoor run to natural enclosures: Changes in behavior in a *Lemur fulvus* troop. *American Journal of Primatology, 4*, 339.

Macedonia, J. M. (1987). Effects of housing differences upon activity budgets in captive sifakas (*Propithecus verreauxi*). *Zoo Biology, 6*, 55–67.

Miller, L. C., Bard, K. A., Juno, C. J., & Nadler, R. D. (1986). Behavioral responsiveness of young chimpanzees (*Pan troglodytes*) to a novel environment. *Folia Primatologica, 47*, 128–142.

Nickelson, S. A., & Lockard, J. S. (1978). Ethogram of Celebes monkeys (*Macaca nigra*) in two captive habitats. *Primates. 19*, 437–447.

Ogden, J. J., Finlay, T. W., & Maple, T. L. (1990). Gorilla adaptations to naturalistic environments. *Zoo Biology, 9*, 107–122.

Raymond, G., & White, L. (1978). Behavioral responses of chimpanzees to behavioral enrichment. *Animal Keepers Forum, 5*, 135–136.

Rowell, T. E. (1974). A quantitative comparison of the behaviour of a wild and a caged baboon group. *Animal Behaviour, 15*, 499–509.

Schoenfeld, D. (1989). Effects of environmental impoverishment on the social behavior of marmosets (*Callithrix jacchus*). *American Journal of Primatology, Suppl. 1*, 45–52.

Tollman, S. G. (1981). Adaptive strategies adopted by a free-ranging troop of vervets when placed in a specially designed enclosed environment. In A. B. Chiarelli & R. S. Corruccini (Eds.), *Primate behavior and sociobiology* (pp. 43–51). New York: Springer-Verlag.

de Waal, F. B. M. (1985). A successful experiment with a natural-sized chimpanzee colony. *Zoonooz, 58*, 4–10.

White, L. (1978). Behavioral response of orangutans to remodeling of their exhibit. *Animal Keepers Forum, 5*, 179–180.

Cage Size (see also Change in Environment):

Alexander, B. K., & Roth, E. M. (1971). The effects of acute crowding on aggressive behavior of Japanese monkeys. *Behaviour, 39*, 73–89.

Anderson, B., Erwin, N., Flynn, D., Lewis, L., & Erwin, J. (1977). Effects of short-term crowding on aggression in captive groups of pigtail monkeys (*Macaca nemestrina*). *Aggressive Behavior, 3*, 33–46.

Clarke, A. S., Juno, C. J., & Maple, T. L. (1982). Behavioral effects of a change in the physical environment: A pilot study of captive chimpanzees. *Zoo Biology, 1*, 371–380.

Dashback, N. J., Schein, M. W., & Haines, D. E. (1982). Cage-size effects on locomotor, grooming and agonistic behaviours of the slow loris, *Nycticebus coucang* (Primates, Lorisidae). *Applied Animal Ethology, 9*, 317–330.

Draper, W. A., & Bernstein, I. W. (1963). Stereotyped behavior and cage size. *Perceptual Motor Skills, 16*, 231–234.

Drickamer, L. C. (1975). Patterns of space utilization and group interactions among free-ranging Macaca mulatta. *Primates. 16*, 23–33.

Elton, R. H. (1979). Baboon behavior under crowded conditions. In J. Erwin, T. L. Maple, & G. Mitchell (Eds.), *Captivity and behavior* (pp. 125–138). New York: Van Nostrand Reinhold.

Elton, R. H., & Anderson, B. V. (1977). The social behavior of a group of baboons (*Papio anubis*) under artificial crowding. *Primates, 18*, 225–234.

Erwin, J. (1977). Factors influencing aggressive behavior and risk of trauma in pigtail monkeys (*Macaca nemestrina*). *Laboratory Animal Science, 27*, 541–547.

Erwin, J. (1979). Aggression in captive macaques: Interaction of social and spatial factors. In J. Erwin, T. L. Maple, & G. Mitchell (Eds.), *Captivity and behavior* (pp. 139–171). New York: Van Nostrand Reinhold.

Erwin, J. (1986). Environments for the captive propagation of primates: Interaction of social and physical factors. In K. Benirschke (Ed.), *Primates: The Road to self-sustaining populations* (pp. 297–306). New York: Springer-Verlag.

Erwin, J., & Deni, R. (1979). Strangers in a strange land: Abnormal behaviors or abnormal environments? In J. Erwin, T. L. Maple, & G. Mitchell (Eds.), *Captivity and behavior* (pp. 1–28). New York: Van Nostrand Reinhold.

Erwin, N., & Erwin, J. (1976). Social density and aggression in captive groups of pigtail monkeys (*Macaca nemestrina*). *Applied Animal Ethology, 2*, 265–269.

Erwin, J., & Sackett, G. P. (1990). Effects of management methods, social organization, and physical space on primate behavior and health. *American Journal of Primatology, 20*, 23–30.

Lemmon, W. B. (1970). Deprivation and enrichment in the development of primates. *Proceedings of the Third International Congress of Primatology, 3*, 108–115.

Line, S. W., Morgan, K. N., Markowitz, H., & Strong, S. (1990). Increased cage size does not alter heart rate or behavior in female rhesus monkeys. *American Journal of Primatology, 20*, 107–114.

Line, S. W., Morgan, K. N., Markowitz, H., & Strong, S. (1989). Influence of cage size on heart rate and behavior in rhesus monkeys. *American Journal of Veterinary Research, 50*, 1523–1526.

McCusker, J. S. (1978). Exhibit size as related to the breeding potential of select primates and carnivores in zoos of the British Isles. *American Association of Zoological Parks and Aquariums Conference Proceedings*, 168–172.

McGuire, M. T., Cole, S. R., & Crookshank, C. (1978). Effects of social and spatial density changes in *Cercopithecus aethiops sabaeus*. *Primates, 19*, 615–632.

Nash, L. T., & Chilton, S. (1986). Space or novelty? Effects of altered cage size on *Galago* behavior. *American Journal of Primatology, 10*, 37–49.

Niewenhuijsen, K., de Waal, F. B. M. (1982). Effects of spatial crowding on social behavior in a chimpanzee colony. *Zoo Biology, 1*, 5–28.

Novak, M. A., & Drewsen, K. D. (1989). Enriching the lives of captive primates: Issues and problems. In E. F. Segal (Ed.), *Housing, care, and psychological wellbeing of captive and laboratory primates* (pp. 161–182). Park Ridge, NJ: Noyes.

Paulk, H. H., Dienske, H., & Ribbens, L. G. (1977). Abnormal behavior in relation to cage size in rhesus monkeys. *Journal of Abnormal Psychology, 86*, 87–92.

Traylor-Holzer, K., & Fritz, P. (1985). Utilization of space by adult and juvenile groups of chimpanzees (*Pan troglodytes*). *Zoo Biology, 4*, 115–127.

Williams, C. J., Mellen, J. D., & King, N. E. (1984). Behavior of a group of captive orangutans (*Pongo pygmaeus abelii*) in two different enclosures. *American Association of Zoological Parks and Aquariums Regional Conference Proceedings*, 186–202.

Wilson, S. F. (1982). Environmental influences on the activity of captive apes. *Zoo Biology, 1*, 201–209.

Enrichment Manipulations:

Visual Exploration

Butler, R. A. (1954). Incentive conditions which influence visual exploration. *Journal of Experimental Psychology, 48*, 19–23.

Butler, R. A. (1957). The effect of deprivation of visual incentives on visual exploration motivation in monkeys. *Journal of Comparative and Physiological Psychology, 50*, 177–179.

Butler, R. A. (1958). The differential effect of visual and auditory incentives on the performance of monkeys. *American Journal of Psychology, 71*, 591–593.

Butler, R. A., & Alexander, H. M. (1955). Daily patterns of visual exploratory behavior in the monkey. *Journal of Comparative and Physiological Psychology, 48*, 247–249.

Haude, R. H., & Ray, O. S. (1967). Visual exploration in monkeys as a function of visual incentive duration and sensory deprivation. *Journal of Comparative and Physiological Psychology, 64*, 332–336.

Haude, R. H., & Ray, O. S. (1974). Visual observing behavior in the squirrel monkey. *Animal Learning and Behavior, 2*, 138–140.

Humphrey, N. K. (1972). "Interest" and "pleasure": Two determinants of a monkey's visual preferences. *Perception, 1*, 395–416.

Roy, M. A., Hennig, C. W., Hill, E. M., & Walker, M. L. (1981). Subspecies differences in curiosity and general activity for developing squirrel monkeys. *Developmental Psychobiology, 4*, 55–58.

Wilcoxon, H. C., Meier, G. W., Orlando, R., & Paulson, D. G. (1969). Visual self-stimulation in socially-living rhesus monkeys. *Proceedings of the Second International Congress of Primatology, 1*, 261–266.

Novel Objects

Baldwin, D. V., & Suomi, S. J. (1974). Reactions of infant monkeys to social and nonsocial stimuli. *Folia Primatologica, 22*, 307–314.

Bayne, K. A. (1989). Nylon balls revisited. *Laboratory Primate Newsletter, 28*, 5–6.

Bloom, K. R., & Cook, M. (1989). Environmental enrichment: Behavioral responses of rhesus to puzzle feeders. *Lab Animal, 18*, 25–31.

Borrebach, D. E. (1977). Two inexpensive manipulanda for use with monkeys or children. *Behavior Research Methods and Instrumentation, 9*, 391–392.

Bryant, C. E., Rupniak, N. M. J., & Iverson, S. D. (1988). Effects of different environmental enrichment devices on cage stereotypies and autoaggression in captive cynomolgous monkeys. *Journal of Medical Primatology, 17*, 257–269.

Chamove, A. S. (1983). The role of dominance in macaque response to novel objects. *Motivation and Emotion, 7*, 213–228.

Champoux, M., Hempel, M., & Reinhardt, V. (1987). Environmental enrichment with sticks for singly caged adult rhesus monkeys. *Laboratory Primate Newsletter, 26*, 5–7.

Crockett, C., Bielitzki, J., Carey, A., & Velez, A. (1989). Kong toys as enrichment devices for singly-caged macaques. *Laboratory Primate Newsletter, 28*, 3–5.

Eterovic, V. A., & Ferchmin, P. A. (1985). Interaction with objects by group-living *Saimiri sciureus*. *Puerto Rico Health Sciences Journal, 4*, 121–125.

Fragaszy, D. M., & Mason, W. A. (1978). Response to novelty in Saimiri and Callicebus: Influence of social context. *Primates, 19*, 311–331.

Gisolfi, C. V., Mora, F., Natterman, R., & Myers, R. D. (1978). New apparatus for exercising a monkey seated in a primate chair. *Journal of Applied Physiology, 44*, 129–132.

Glick, C. (1977). Inexpensive recreational opportunities for a gorilla. *Animal Keepers Forum, 4*, 140.

Line, S. W., Clarke, A. S., & Markowitz, H. (1989). Adult female rhesus macaque responses to novel objects. *Lab Animal, 18*, 33–35, 37, & 39–40.

Maki, S., & Bloomsmith, M. A. (1989). Uprooted trees facilitate the psychological well-being of captive chimpanzees. *Zoo Biology, 8*, 79–88.

Menzel, E. W. (1969). Chimpanzee utilization of space and responsiveness to objects: Age difference and comparison with macaques. *Proceedings of the Second International Congress of Primatology, 1*, 72–80.

Menzel, E. W. (1971). Group behavior in young chimpanzees: Responsiveness to cumulative novel changes in a large outdoor enclosure. *Journal of Physiology and Psychobiology, 74*, 46–51.

Menzel, E. W., & Menzel, C. R. (1979). Cognitive, developmental, and social aspects of responsiveness to novel objects in a family group of marmosets (*Saguinus fuscicollis*). *Behaviour, 70*, 251–279.

Millar, S. K., Evans, S., & Chamove, A. S. (1988). Older offspring contact novel objects soonest in Callitrichid families. *Biology of Behavior, 13*, 82–96.

Murphy, D. E. (1976). Enrichment and occupational devices for orangutans and chimpanzees. *International Zoo News, 23*, 24–26

Paquette, D., & Prescott, J. (1988). Use of novel objects to enhance environments of captive chimpanzees. *Zoo Biology, 7*, 15–24.

Preilowski, B., Reger, M., & Engele, H. (1988). Combining scientific experimentation with conventional housing: A pilot study with rhesus monkeys. *American Journal of Primatology, 14*, 223–234.

Reinhardt, V. (1987). Improved installation method for branches as cage enrichment. *Laboratory Primate Newsletter, 26*, 1–3.

Reinhardt, V. (1990). Time budget of caged rhesus monkeys exposed to a companion, a PVC perch, and a piece of wood for an extended time. *American Journal of Primatology, 20*, 51–56.

Reinhardt, V., Houser, W. D., Cowley, D., & Champoux, M. (1987). Preliminary comments on environmental enrichment with branches for individually caged rhesus monkeys. *Laboratory Primate Newsletter, 26*, 1–3.

Reinhardt, V., & Smith, M. D. (1988). PVC pipes effectively enrich the environment of caged rhesus monkeys. *Laboratory Primate Newsletter, 27*, 4–5.

Renquist, D. M., & Judge, F. J. (1985). Use of Nylaballs as behavioral modifiers for caged primates. *Laboratory Primate Newsletter, 24*, 4.

Ross, P. W., & Everitt, J. I. (1988). A nylon ball device for primate environmental enrichment. *Laboratory Animal Science, 38*, 481–483.

Torigoe, T. (1985). Comparison of object manipulation among 74 species of nonhuman primates. *Primates, 26*, 182–194.

Tripp, J. K. (1985). Increasing activity in captive orangutans: Provision of manipulable and edible materials. *Zoo Biology, 4*, 225–234.

Vauclair, J., & Bard, K. A. (1983). Development of manipulations with objects in ape and human infants. *Journal of Human Evolution, 12*, 631–645.

Visalberghi, E. (1988). Responsiveness to objects in two social groups of tufted capuchin monkeys (*Cebus apella*). *American Journal of Primatology, 15*, 349–360.

Watson, D. S. B., Houston, B. J., & Macallum, G. E. (1989). The use of toys for primate environmental enrichment. *Laboratory Primate Newsletter, 28*, 20.

Westergaard, G. C., & Fragaszy, D. M. (1985). Effects of manipulatable objects on the activity of captive capuchin monkeys (*Cebus apella*). *Zoo Biology, 4*, 317–327.

Wilson, S. F. (1982). Environmental influences on the activity of captive apes. *Zoo Biology, 1*, 201–209.

Wolff, A. V. (1989). Polyvinyl chloride piping as perch material for squirrel monkeys. *Laboratory Primate Newsletter, 28*, 7.

Novel Objects and Tool Use

Antinucci, F., & Visalberghi, E. (1986). Tool use in *Cebus apella*: A case study. *International Journal of Primatology, 7*, 349–361.

Beck, B. B. (1980). *Animal tool behavior: The use and manufacture of tools by animals*. New York: Garland STPM Press.

Fragaszy, D. M., & Visalberghi, E. (1989). Social influences on the acquisition of tool-using behaviors in tufted capuchin monkeys (*Cebus apella*). *Journal of Comparative Psychology, 103*, 159–170.

Maki, S., Alford, P. L., Bloomsmith, M. A., & Franklin, J. (1989). Food puzzle device simulating termite fishing for captive chimpanzees (*Pan troglodytes*). *American Journal of Primatology, Suppl. 1*, 71–78.

Menzel, E. W. (1972). Spontaneous invention of ladders in a group of young chimpanzees. *Folia Primatologica, 17*, 87–106.

Menzel, E. W., Davenport, R. K., & Rogers, C. M. (1970). The development of tool-using in wild-born and restriction-reared chimpanzees. *Folia Primatologica, 12*, 273–283.

Nash, V. J. (1982). Tool use by captive chimpanzees at an artificial termite mound. *Zoo Biology, 1*, 211–221.

Visalberghi, E. (1986). Acquisition of nut-cracking behavior by two capuchin monkeys (*Cebus apella*). *Primate Report, 14*, 226–227.

Visalberghi, E., & Antinucci, F. (1986). Tool use in the exploitation of food resources in *Cebus apella*. In J. G. Else & P. C. Lee (Eds.), *Primate ecology and conservation* (Vol. 2, pp. 57–62). Cambridge: Cambridge University Press.

Visalberghi, E., & Vitale, A. F. (1990). Coated nuts as an enrichment device to elicit tool use in tufted capuchins (*Cebus apella*). *Zoo Biology, 9*, 65–72.

Westergaard, G. C. (1988). Lion-tailed macaques (*Macaca silenus*) manufacture and use tools. *Journal of Comparative Psychology, 102*, 152–159.

Westergaard, G. C., & Fragaszy, D. M. (1987). The manufacture and use of tools by capuchin monkeys (*Cebus apella*). *Journal of Comparative Psychology, 2*, 159–168.

Westergaard, G. C., & Lindquist, T. (1987). Manipulation of objects in a captive group of lion-tailed macaques (*Macaca silenus*). *American Journal of Primatology, 12*, 231–234.

Foraging Demands, Tasks, and Devices

Anderson, J. R., & Chamove, A. S. (1984). Allowing captive primates to forage. In *Standards in Laboratory Animal Management* (Part 2, pp. 253–256). Potters Bar, UK: Universities Federation for Animal Welfare.

Beckley, S., & Novak, M. (1989). An examination of various foraging components and their suitability as enrichment tools for captively housed primates. *American Journal of Primatology, Suppl. 1*, 37–44.

Bloomsmith, M. A. (1989). Feeding enrichment for captive great apes. In E. F. Segal (Ed.), *Housing, care, and psychological wellbeing of captive and laboratory primates* (pp. 336–356). Park Ridge, NJ: Noyes.

Bloomsmith, M. A., Alford, P. L., & Maple, T. L. (1988). Successful feeding enrichment for captive chimpanzees. *American Journal of Primatology, 16*, 155–164.

Bloomstrand, M., Riddle, K., Alford, P., & Maple, T. L. (1986). Objective evaluation of a behavioral enrichment device for captive chimpanzees (*Pan troglodytes*). *Zoo Biology, 5*, 293–300.

Boccia, M. L. (1989). Preliminary report on the use of a natural foraging task to reduce aggression and stereotypies in socially housed pigtail macaques. *Laboratory Primate Newsletter, 28*, 3–4.

Boccia, M. L. (1989). Long-term effects of a natural foraging task on aggression and stereotypies in socially housed pigtail macaques. *Laboratory Primate Newsletter, 28*, 18–19.

Boccia, M. L., Laudenslager, M., & Reite, M. (1988). Food distribution, dominance, and aggressive behaviors in bonnet macaques. *American Journal of Primatology, 16*, 123–130.

Chamove, A. S., & Anderson, J. R. (1988). Impact of feeding practice on growth and behavior of stumptailed macaques (*Macaca arctoides*). In J. E. Fa & C. H. Southwick (Eds.), *Ecology and behavior of food-enhanced primate groups* (pp. 231–246). New York: Alan R. Liss.

Gust, D. A. (1989). Uncertain availability of a preferred food affects choice in a captive group of chimpanzees. *American Journal of Primatology, 17*, 165–172.

Gust, D. A., Swenson, R. B., Smith, M. B., & Sikes, J. (1988). An apparatus and procedure for studying the choice component of foraging in a captive group of chimpanzees. *Primates, 29*, 139–143.

Markowitz, H. (1979). Environmental enrichment and behavioral engineering for captive primates. In J. Erwin, T. L. Maple, & G. Mitchell (Eds.), *Captivity and behavior* (pp. 217–238). New York: Van Nostrand Reinhold Co.

Marriott, B. M., & Salzen, E. A. (1979). Food storing behavior in captive squirrel monkeys (*Saimiri sciureus*). *Primates, 20*, 307–311.

McGrew, W. C., Brennan, J. A., & Russell, J. (1986). An artificial "gum-tree" for marmosets (*Callithrix j. jacchus*). *Zoo Biology, 5*, 45–50.

Menzel, E. W., & Juno, C. (1982). Social foraging in marmoset monkeys and the question of intelligence. *Philosophical Transactions of the Royal Society of London* [*Biology*], *308*, 145–158.

Moazed, T. C., & Wolff, A. V. (1988). The raisin board as an environmental enrichment tool for laboratory primates. *Laboratory Primate Newsletter, 27*, 16.

Molzen, E. M., & French, J. A. (1989). The problem of foraging in captive callitrichid primates: Behavioral time budgets and foraging skills. In E. F. Segal (Ed.), *Housing, care, and psychological wellbeing of captive and laboratory primates* (pp. 89–101). Park Ridge, NJ: Noyes.

Murphy, D. E. (1976). Enrichment and occupational devices for orangutans and chimpanzees. *International Zoo News, 23*, 24–26.

Petto, A. J., & Devin, M. (1988). Food choices in captive common marmosets (*Callithrix jacchus*). *Laboratory Primate Newsletter, 27*, 7–9.

Plimpton, E. H., Swartz, K. B., & Rosenblum, L. A. (1981). The effects of a foraging demand on social interactions in a laboratory group of bonnet macaques. *International Journal of Primatology, 2*, 177–187.

Rosenblum, L. A., & Paully, G. S. (1984). The effects of varying environmental demands on maternal and infant behavior. *Child Development, 55*, 305–314.

Rosenblum, L. A., & Smiley, J. (1984). Therapeutic effects of an imposed foraging task in disturbed monkeys. *Journal of Child Psychology and Psychiatry, 25*, 485–497.

Smith, A., Lindburg, D. G., & Vehrencamp, S. (1989). Effect of food preparation on feeding behavior of lion-tailed macaques. *Zoo Biology, 8*, 57–66.

de Waal, F. B. M. (1984). Coping with social tension: Sex differences in the effect of food provision to small rhesus monkey groups. *Animal Behaviour, 32*, 765–773.

Yanofsky, R., & Markowitz, H. (1978). Changes in general behavior of two mandrills (*Papio sphinx*) concomitant with behavioral testing in the zoo. *Psychological Record, 28*, 369–373.

Mirror Stimulation

Anderson, J. R. (1983). Mirror-image stimulation and short separations in stumptail monkeys. *Animal Learning and Behavior, 11*, 139–143.

Anderson, J. R. (1986). Infant stumptailed macaques reared with mirrors or peers: Social responsiveness, attachment, and adjustment. *Primates, 27*, 63–82.

Anderson, J. R. (1986). Mirror-mediated finding of hidden food by monkeys (*Macaca tonkeana* and *M. fascicularis*). *Journal of Comparative Psychology, 100*, 237–242.

Bayart, F. (1985). Mirror-image reactions in a tool using, adult male *Macaca tonkeana*. *Behavioral Processes, 10*, 219–227.

Collinge, N. E. (1989). Mirror reactions in a zoo colony of Cebus monkeys. *Zoo Biology, 8*, 89–98.

Eglash, A. R., & Snowdon, C. T. (1983). Mirror-image responses in pygmy marmosets (*Cebuella pygmaea*). *American Journal of Primatology, 5*, 211–220.

Hall, K. R. (1962). Behavior of monkeys towards mirror images. *Nature, 196*, 1258–1261.

Itakura, S. (1987). Use of a mirror to direct their responses in Japanese monkeys (*Macaca fuscata*). *Primates, 28*, 343–352.

Platt, M. M., & Thompson, R. L. (1985). Mirror responses in Japanese macaque troop (Arashiyama West). *Primates, 26*, 300–314.

Robert, S. (1986). Ontogeny of mirror behavior in two species of great apes. *American Journal of Primatology, 10*, 109–117.

Suarez, S., & Gallup, G. (1986). Social responding to mirrors in rhesus macaques (*Macaca mulatta*): Effects of changing mirror location. *American Journal of Primatology, 11*, 239–244.

Videotapes, Television, Photographs, Videogames

Brent, L., Lee, D. R., & Eichberg, J. W. (1989). Evaluation of two enrichment devices for singly caged chimpanzees (*Pan troglodytes*). *American Journal of Primatology, Suppl. 1*, 65–70.

Demaria, C., & Thierry, B. (1988). Responses to animal stimulus photographs in stumptailed macaques. *Primates, 29*, 237–244.

Rumbaugh, D. M., Richardson, W. K., Washburn, D. A., Savage-Rumbaugh, E. S., & Hopkins, W. D. (1989). Rhesus monkeys (*Macaca mulatta*), video tasks, and implications for stimulus-response spatial contiguity. *Journal of Comparative Psychology, 103*, 32–38.

Sackett, G. P. (1965). Response of rhesus monkeys to social stimulation presented by means of colored slides. *Perceptual and Motor Skills, 20*, 1027–1028.

Swartz, K. B., & Rosenblum, L. A. (1980). Operant responding by bonnet macaques for color videotape recordings of social stimuli. *Animal Learning and Behavior, 8*, 311–321.

Other Manipulations

Brown, S. G., Dunlap, W. P., & Maple, T. L. (1982). Notes on water contact by a captive male lowland gorilla. *Zoo Biology, 1*, 243–249.

Haraway, M., Maples, E., Tolson, S., & Yelverton, J. (1984). Environmental support of vocal behavior in captive gibbons. *American Association of Zoological Parks and Aquariums Regional Proceedings*, 301–313.

Control Over Environment

Hanson, J. P., Larson, M. E., & Snowdon, C. T. (1976). The effects of control over high intensity noise on plasma cortisol levels in rhesus monkeys. *Behavioral Biology, 16*, 333–340.

Mineka, S., Gunnar, M., & Champoux, M. (1986). Control and early socioemotional development: Infant rhesus monkeys reared in controllable versus uncontrollable environments. *Child Development, 57*, 1241–1256.

SOCIAL ENVIRONMENT

An examination of those variables such as rearing experiences, group versus individual housing, group composition, and group formation that may affect the welfare of nonhuman primates.

Role of Early Experience:

General Articles

Capitanio, J. P. (1986). Behavioral pathology. In G. Mitchell & J. Erwin (Eds.), *Comparative primate biology: Vol. 2A. Behavior, conservation, and ecology* (pp. 411–454). New York: Alan R. Liss.

Erwin, J., & Deni, R. (1979). Strangers in a strange land: Abnormal behaviors or abnormal environments? In J. Erwin, T. L. Maple, & G. Mitchell (Eds.), *Captivity and behavior* (pp. 1–28). New York: Van Nostrand Reinhold.

Harlow, H. F., & Harlow, M. K. (1965). The affectional systems. In A. M. Schrier, H. F. Harlow, & F. Stollnitz (Eds.), *Behavior of nonhuman primates* (Vol. 2, pp. 287–334). New York: Academic Press.

Hinde, R. A. (1971). Development of social behavior. In A. M. Schrier & F. Stollnitz (Eds.), *Behavior of nonhuman primates* (Vol. 3, pp. 1–68). New York: Academic Press.

Mason, W. A. (1965). The social development of monkeys and apes. In I. DeVore (Ed.), *Primate behavior: Field studies of monkeys and apes* (pp. 514–543). New York: Holt.

Mason, W. A. (1979). Ontogeny of social behavior. In P. Marler & J. G. Vandenbergh (Eds.), *Handbook of behavioral neurobiology: Vol. 3. Social behavior and communication* (pp. 1–28). New York: Plenum Press.

Mitchell, G. (1970). Abnormal behavior in primates. In L. A. Rosenblum (Ed.), *Primate behavior: Developments in field and laboratory research* (Vol. 1, pp. 196–249). New York: Academic Press.

Ruppenthal, G. C., & Sackett, G. P. (1979). Experimental and husbandry procedures: Their impact on development. In G. C. Ruppenthal & G. P. Sackett (Eds.), *Nursery care of nonhuman primates* (pp. 269–284). New York: Plenum Press.

Sackett, G. P. (1965). Effect of rearing conditions upon monkeys. *Child Development, 35*, 855–868.

Sackett, G. P. (1970). Innate mechanisms, rearing conditions, and a theory of early experience effects in primates. In M. R. Jones (Ed.), *Miami symposium on prediction of behavior: Early experience* (pp. 1–45). Miami: University of Miami Press.

Isolation Rearing

Animals reared in isolation from conspecifics during the first year of life often develop bizarre behavior patterns and have low reproductive fitness.

Anderson, C. O., & Mason, W. A. (1974). Early experience and complexity of social organization in groups of young rhesus monkeys (*Macaca mulatta*). *Journal of Comparative and Physiological Psychology, 87*, 681–690.

Anderson, J. R., & Chamove, A. S. (1980). Self-aggression and social aggression in laboratory reared macaques. *Journal of Abnormal Psychology, 89*, 539–550.

Anderson, J. R., & Chamove, A. S. (1981). Self-aggressive behaviour in monkeys. *Current Psychological Reviews, 1*, 139–158.

Arling, G. L., & Harlow, H. F. (1967). Effects of social deprivation on maternal behavior of rhesus monkeys. *Journal of Comparative and Physiological Psychology, 64*, 371–377.

Bajpai, R. K. (1980). The effect of early social deprivation on rhesus monkey social responsiveness. *Primates, 21*, 510–519.

Berkson, G. (1968). Development of abnormal stereotyped behaviors. *Developmental Psychobiology, 1*, 118–132.

Berkson, G., & Mason, W. A. (1964). Stereotyped behaviors of chimpanzees: Relation to general arousal and alternate activities. *Perceptual Motor Skills, 19*, 635–652.

Brandt, E. M., Stevens, C. W., & Mitchell, G. (1971). Visual social communication in adult male isolate-reared monkeys (*Macaca mulatta*). *Primates, 12*, 105–112.

Chamove, A. S., & Anderson, J. R. (1981). Self-aggression, stereotypy, and self-injurious behaviour in man and monkeys. *Current Psychological Reviews, 1*, 245–256.

Coelho, A. M., & Bramblett, C. A. (1981). Effects of rearing on aggression and subordination in papio monkeys. *American Journal of Primatology, 1*, 401–412.

Cross, H. A., & Harlow, H. F. (1965). Prolonged and progressive effects of partial isolation on the behavior of macaque monkeys. *Journal of Experimental Research in Personality, 1*, 39–49.

Davenport, R. K., Rogers, C. M., & Rumbaugh, D. M. (1973). Long-term cognitive deficits in chimpanzees associated with early impoverished rearing. *Developmental Psychology, 9*, 343–347.

Dienske, H., & de Jonge, G. (1982). A comparison of development and deprivation in non-human primates and man. *Journal of Human Evolution, 11*, 511–516.

Dienske, H., & Griffin, R. (1978). Abnormal behavior patterns developing in chimpanzee infants during nursery care: A note. *Journal of Child Psychology and Psychiatry, 19*, 387–391.

Elias, M. F., & Sammonds, K. W. (1973). Exploratory behavior of cebus monkeys after having been reared in partial isolation. *Child Development, 44*, 218–220.

Faucheux, B., Bertrand, M., & Bourliere, F. (1978). Some effects of living conditions upon the pattern of growth in the stumptail macaque. *Folia Primatologica, 30*, 220–236.

Fittinghoff, N. A., Lindburg, D. G., Gomber, J., & Mitchell, G. (1974). Consistency and variability in the behavior of mature isolation-reared male rhesus macaques. *Primates, 15*, 111–139.

Gluck, J. P., Harlow, H. F., & Schiltz, K. A. (1973). Differential effect of early enrichment and deprivation on learning in the rhesus monkey (*Macaca mulatta*). *Journal of Comparative and Physiological Psychology, 84*, 598–604.

Goosen, C. (1981). Abnormal behaviour patterns in rhesus monkeys: Symptoms of mental disease? *Biological Psychiatry, 16*, 697–716.

Goy, R. W., Wallen, K., & Goldfoot, D. A. (1978). Social factors affecting the development of mounting behavior in male rhesus monkeys. In W. Montagna & W. A. Sadler (Eds.), *Reproductive behavior* (pp. 223–247). New York: Plenum Press.

Goy, R. W., & Wallen, K. (1979). Experiential variables influencing play, foot clasp mounting and adult sexual competence in male rhesus monkeys. *Psychoneuroendocrinology, 4*, 1–12.

Griffin, G. A., & Harlow, H. F. (1966). Effects of three months of total social deprivation on the social adjustment and learning in the rhesus monkey. *Child Development, 37*, 533–547.

Harlow, H. F. (1958). The nature of love. *American Psychologist, 13*, 673–685.

Harlow, H. F., Dodsworth, R. O., & Harlow, M. K. (1965). Total social isolation in monkeys. *Proceedings of the National Academy of Science, 54*, 90–97.

Harlow, H. F., & Harlow, M. K. (1962). Social deprivation in monkeys. *Scientific American, 207*, 137–146.

Harlow, H. F., Harlow, M. K., Dodsworth, R. O., & Arling, G. L. (1966). Maternal behavior of rhesus monkeys deprived of mothering and peer associations in infancy. *Proceedings of the American Philosophical Society, 110*, 58–66.

Harlow, H. F., Rowland, G. L., & Griffin, G. A. (1964). The effect of total social deprivation on the development of monkey behavior. *Psychiatric Research Reports, 9*, 116–135.

Harlow, H. F., Schiltz, K. A., & Harlow, M. K. (1969). Effects of social isolation on the learning performance of rhesus monkeys. *Proceedings of the Second International Congress of Primatology, 1*, 178–185.

Harlow, H. F., & Seay, B. (1966). Mothering in motherless mother monkeys. *British Journal of Social Psychiatry, 1*, 63–69.

Herzog, H., & Hopf, S. (1984). Behavioral responses to species-specific warning calls in infant squirrel monkeys reared in social isolation. *American Journal of Primatology, 7*, 99–106.

Kerr, G. R., Chamove, A. S., & Harlow, H. F. (1969). Environmental deprivation: Its effects on the growth of infant monkeys. *Journal of Pediatrics, 75*, 833–837.

Lichstein, L., & Sackett, G. P. (1971). Reactions by differentially raised rhesus monkeys to noxious stimulation. *Developmental Psychobiology, 4*, 339–352.

Mason, W. A. (1961). Effects of age and stimulus characteristics on manipulatory responsiveness of monkeys raised in a restricted environment. *Journal of Genetic Psychology, 99*, 301–308.

Mason, W. A. (1961). The effects of social restriction on the behavior of rhesus monkeys: Tests of gregariousness. *Journal of Comparative and Physiological Psychology, 54*, 287–290.

Mason, W. A. (1961). The effects of social restriction on the behavior of rhesus monkeys: III. Dominance tests. *Journal of Comparative and Physiological Psychology, 54*, 694–699.

Mason, W. A. (1985). Experiential influences on the development of expressive behaviors in rhesus monkeys. In G. Zivin (Ed.), *The development of ex-*

pressive behavior: Biology–environment interactions (pp. 117–152). New York: Academic Press.

Mason, W. A., & Green, P. C. (1962). The effects of social restriction on the behavior of rhesus monkeys: IV. Responses to a novel environment and an alien species. *Journal of Comparative and Physiological Psychology, 55*, 363–368.

McKinney, W. T. (1974). Primate social isolation. *Archives of General Psychiatry, 31*, 422–426.

Meyer, J. S., & Bowman, R. E. (1972). Rearing experience, stress and adrenocorticosteroids in the rhesus monkey. *Physiology and Behavior, 8*, 339–343.

Miller, R. E., Caul, W. F., & Mirsky, I. A. (1971). Patterns of eating and drinking in socially-isolated rhesus monkeys. *Physiology and Behavior, 7*, 127–134.

Miller, R. E., Mirsky, I. A., Caul, W. F., & Sakata, T. (1969). Hyperphagia and polydipsia in socially-isolated rhesus monkeys. *Science, 165*, 1027–1028.

Mitchell, G., Raymond, E. J., Ruppenthal, G. C., & Harlow, H. F. (1966). Long-term effects of total social isolation upon behavior of rhesus monkeys. *Psychological Reports, 18*, 567–580.

Nadler, R. D. (1980). Child abuse: Evidence from nonhuman primates. *Developmental Psychobiology, 13*, 507–512.

Newman, J. D., & Symmes, D. (1974). Vocal pathology in socially deprived monkeys. *Developmental Psychobiology, 7*, 351–358.

Reite, M., & Short, R. (1977). Nocturnal sleep in isolation-reared monkeys: Evidence for environmental independence. *Developmental Psychobiology, 10*, 555–561.

Roy, M. A. (1981). Abnormal behaviors in nursery-reared squirrel monkeys (*Saimiri sciureus*). *American Journal of Primatology, 1*, 35–42.

Sackett, G. P. (1965). Manipulatory behavior in monkeys reared under different levels of early stimulus variation. *Perceptual Motor Skills, 20*, 985–988.

Sackett, G. P. (1965). Response of rhesus monkeys to social stimulation presented by means of colored slides. *Perceptual Motor Skills, 20*, 1027–1028.

Sackett, G. P. (1966). Monkeys reared in isolation with pictures as visual input: Evidence for an innate releasing mechanism. *Science, 154*, 1468–1472.

Sackett, G. P. (1968). The persistence of abnormal behaviour in monkeys following isolation rearing. In R. Porter (Ed.), *Ciba foundation symposium: The role of learning in psychotherapy* (pp. 3–25). London: J. & A. Churchill Ltd.

Sackett, G. P. (1972). Exploratory behavior of rhesus monkeys as a function of rearing experiences and sex. *Developmental Psychobiology, 6*, 260–270.

Sackett, G. P., Bowman, R. E., Meyer, J. S., Tripp, R. L., & Grady, S. S. (1973). Adrenocortical and behavioral reactions by differentially raised rhesus monkeys. *Physiological Psychology, 1*, 209–212.

Sackett, G. P., Holm, R. A., & Ruppenthal, G. C. (1976). Social isolation rearing: Species differences in behavior of macaque monkeys. *Developmental Psychology, 12*, 283–288.

Sackett, G. P., Porter, M., & Holmes, H. (1965). Choice behavior in rhesus monkeys: Effect of stimulation during the first month of life. *Science, 147*, 304–306.

Seay, B., Alexander, B. K., & Harlow, H. F. (1964). Maternal behavior of socially deprived rhesus monkeys. *Journal of Abnormal and Social Psychology, 69*, 345–354.

Stroebel, C. F. (1969). Biological rhythm correlates of disturbed behavior in the rhesus monkey. *Bibliotheca Primatologica, 9*, 91–105.

Struble, R. G., & Riesen, A. H. (1977). Changes in cortical dendritic branching subsequent to partial social isolation in stumptailed monkeys. *Developmental Psychobiology, 11*, 479–486.

Suomi, S. J. (1978). Maternal behavior by socially incompetent monkeys: Neglect and abuse of offspring. *Journal of Pediatric Psychology, 3*, 28–34.

Surrogate Rearing

Animals reared only with inanimate surrogates tend to develop bizarre behavior patterns and have low reproductive fitness.

Ainsworth, L. L., & Baker, C. D. (1982). Effects of illumination changes on infant monkey contacts with surrogates. *Psychological Record, 32*, 513–518.

Anderson, C. O., Kenney, A. M., & Mason, W. A. (1977). Effects of maternal mobility, partner, and endocrine state on social responsiveness of adolescent rhesus monkeys. *Developmental Psychobiology, 10*, 421–434.

Harlow, H. F., & Harlow, M. K. (1966). Learning to love. *American Scientist, 54*, 244–272.

Harlow, H. F., & Suomi, S. J. (1970). Nature of love—simplified. *American Psychologist, 25*, 161–168.

Hill, S. D., McCormack, S. A., & Mason, W. A. (1973). Effects of artificial mothers and visual experience on adrenal responsiveness of infant monkeys. *Developmental Psychobiology, 6*, 421–429.

Kaplan, J. (1974). Growth and behavior of surrogate-reared squirrel monkeys. *Developmental Psychobiology, 7*, 7–13.

Kaplan, J., & Shusterman, R. J. (1972). Social preferences of mother and infant squirrel monkeys following different rearing experiences. *Developmental Psychobiology, 5*, 53–59.

Roy, M. A., Wolf, R. H., Martin, L. N., Rangan, S. R. S., & Allen, W. P. (1978). Social and reproductive behaviors in surrogate-reared squirrel monkeys (*Saimiri sciureus*). *Laboratory Animal Science, 28*, 417–421.

Timmermans, P. J., Roder, E. L., & Kemps, M. M. (1988). Rearing cynomolgus monkeys (*Macaca fascicularis*) on surrogate mothers with bottle feeding. *Laboratory Animals, 22*, 229–234.

Peer Rearing

Animals reared with other naive agemates often develop appropriate social responsiveness and are capable of breeding. Under some circumstances, behavioral development may be delayed and reproductive fitness compromised.

Anderson, J. R. (1986). Infant stumptailed macaques reared with mirrors and peers: Social responsiveness, attachment, and adjustment. *Primates, 27*, 63–82.

Chamove, A. S., Rosenblum, L. A., & Harlow, H. F. (1973). Monkeys raised only with peers: a pilot study. *Animal Behaviour, 21*, 316–325.

Harlow, H. F. (1969). Age-mate or peer affectional systems. In D. S. Lehrman, R. A. Hinde, & E. Shaw (Eds.), *Advances in the study of behavior* (Vol. 2, pp. 333–383). New York: Academic Press.

Hopf, H., & Herzog, M. (1985). Onset of peer interactions in socially inexperienced squirrel monkeys. *Folia Primatologica, 44*, 121–128.

Pratt, C. L., & Sackett, G. P. (1967). Selection of social partners as a function of peer contact during rearing. *Science, 155*, 1133–1135.

Suomi, S. J. (1979). Peers, play, and primary prevention in primates. In M. W. Kent & J. E. Rolf (Eds.), *Primary prevention of psychopathology:* Vol. 3. *Social competence in children* (pp. 127–149). Hanover, NH: Universities Press of New England.

Suomi, S. J., & Harlow, H. F. (1975). The role and reason of peer relationships in rhesus monkeys. In M. Lewis & L. A. Rosenblum (Eds.), *The origins of behavior: Friendship and peer relationships* (pp. 153–183). New York: John Wiley & Sons.

Rearing With Other Species

Capitanio, J. P. (1984). Early experience and social processes in rhesus macaques (*Macaca mulatta*). I. Dyadic social interaction. *Journal of Comparative Psychology, 98*, 35–44.

Mason, W. A., & Capitanio, J. P. (1988). Formation and expression of filial attachment in rhesus monkeys raised with living and inanimate mother substitutes. *Developmental Psychobiology, 21*, 401–430.

Mason, W. A., & Kenney, M. D. (1974). Redirection of filial attachments in rhesus monkeys: Dogs as mother surrogates. *Science, 183*, 1209–1211.

Owren, M. J., & Dieter, J. A. (1989). Infant cross-fostering between Japanese (*Macaca fuscata*) and rhesus macaques (*M. mulatta*). *American Journal of Primatology, 18*, 245–250.

Wood, B. S., Mason, W. A., & Kenney, M. D. (1979). Contrasts in visual responsiveness and emotional arousal between rhesus monkeys raised with living and those raised with inanimate substitute mothers. *Journal of Comparative and Physiological Psychology, 93*, 368–377.

Reversibility of Early Experience Effects

The extent to which primates can recover both social and reproductive functioning following impoverished early rearing.

Chamove, A. S. (1978). Therapy of isolate rhesus: Different partners and social behavior. *Child Development, 49*, 43–50.

Cummins, M. S., & Suomi, S. J. (1976). Long-term effects of social rehabilitation in rhesus monkeys. *Primates, 17*, 43–52.

Dienske, H., Vreeswijk, W., & Koning, H. (1980). Adequate mothering by partially isolated rhesus monkeys after observation of maternal care. *Journal of Abnormal Psychology, 89*, 489–492.

Erwin, J., Maple, T. L., Mitchell, G., & Willott, J. (1974). Follow-up study of isolation-reared and mother-reared rhesus monkeys paired with preadolescent conspecifics in late infancy: Cross-sex pairings. *Developmental Psychology, 10*, 808–814.

Fritz, J. (1989). Resocialization of captive chimpanzees: An amelioration procedure. *American Journal of Primatology, Suppl. 1*, 79–86.

Fritz, P., & Fritz, J. (1979). Resocialization of chimpanzees: Ten years of experience at the Primate Foundation of Arizona. *Journal of Medical Primatology, 8*, 202–221.

Gomber, J., & Mitchell, G. (1974). Preliminary report on adult male isolation-reared rhesus monkeys caged with infants. *Developmental Psychology, 10*, 298.

Hannah, A. C., & Brotman, B. (1990). Procedures for improving maternal behavior in captive chimpanzees. *Zoo Biology, 9*, 233–240.

Harlow, H. F., Harlow, M. K., & Suomi, S. J. (1971). From thought to therapy: Lessons from a primate laboratory. *American Scientist, 59*, 538–549.

Harlow, H. F., & Suomi, S. J. (1971). Social recovery by isolation-reared monkeys. *Proceedings of the National Academy of Science, 68*, 1534–1538.

Huebner, D. K., & King, J. E. (1984). Kittens as therapists: Social behavior sequences in isolated squirrel monkeys after exposure to young nonconspecifics. *Developmental Psychobiology, 17*, 233–242.

Hopf, S. (1981). Conditions of failure and recovery of maternal behavior in captive squirrel monkeys (*Saimiri*). *International Journal of Primatology, 2*, 335–349.

Nankivell, B., Fritz, J., Nash, L., & Fritz, P. (1988). Competent maternal behavior by a hand-reared, resocialized, primiparous chimpanzee: A case history. *Laboratory Primate Newsletter, 27*, 5–8.

Novak, M. A. (1979). Social recovery of monkeys isolated for the first year of life: II. Long-term assessment. *Developmental Psychology, 15*, 50–61.

Novak, M. A., & Harlow, H. F. (1975). Social recovery of monkeys isolated for the first year of life: I. Rehabilitation and therapy. *Developmental Psychology, 11*, 453–465.

Pfeiffer, A. J., & Koebner, L. J. (1978). The resocialization of single-caged chimpanzees and the establishment of an island colony. *Journal of Medical Primatology, 7*, 70–81.

Rijksen, H., & Rijksen, A. (1979). Rehabilitation, a new approach is needed. *Tigerpaper, 6*, 16–18.

Strongin, T. S., Gluck, J. P., & Frank, R. G. (1977). Development of social behavior in an adult total isolate rhesus monkey (*Macaca mulatta*). *Journal of Autism and Childhood Schizophrenia, 7*, 329–336.

Suomi, S. J., & Harlow, H. F. (1972). Social rehabilitation of isolate-reared monkeys. *Developmental Psychology, 6*, 487–496.

Group Formation and Introductions:

Bernstein, I. (1964). The integration of rhesus monkeys introduced to a group. *Folia Primatologica, 2*, 50–63.

Bernstein, I. S. (1964). Group social patterns as influenced by removal and later reintroduction of the dominant male rhesus. *Psychological Reports, 14*, 3–10.

Bernstein, I. S. (1969). Introductory techniques in the formation of pigtail monkey troops. *Folia Primatologica, 10*, 1–19.

Bernstein, I. S. (1971). The influence of introductory techniques on the formation of captive mangabey groups. *Primates, 12*, 33–44.

Bernstein, I. S., Gordon, T. P., & Rose, R. M. (1974). Factors influencing the expression of aggression during introductions to rhesus monkey groups. In R. Holloway (Ed.), *Primate aggression, territoriality and xenophobia* (pp. 211–240). New York: Academic Press.

Bernstein, I. S., Gordon, T. P., & Rose, R. M. (1974). Aggression and social controls in rhesus monkey (*Macaca mulatta*) groups revealed in group formation studies. *Folia Primatologica, 21*, 81–107.

Bernstein, I. S., & Mason, W. A. (1963). Group formation by rhesus monkeys. *Animal Behaviour, 11*, 28–31.

Bernstein, I. S., Rose, R. M., & Gordon, T. P. (1977). Behavioural and hormonal responses of male rhesus monkeys introduced to females in the breeding and nonbreeding seasons. *Animal Behaviour, 25*, 609–614.

Bowen, R. A. (1981). Social integration in lowland gorillas. *Dodo: Jersey Preservation Trust, 18*, 51–59.

Bunnell, B. N., Kenshalo, D. P., Czerny, P., & Allen, J. D. (1979). Performance correlates of social behavior and organization: effects of group formation on operant performance in rhesus monkeys (*Macaca mulatta*). *Primates, 20*, 197–210.

Caine, N. G., & Short, J. (1981). Introducing unfamiliar monkeys (*Macaca nemestrina* and *M. radiata*) to established social groups. *Laboratory Primate Newsletter, 20*, 1–4.

Candland, D. K., O'Connor, K., Dresdale, L., Leshner, A., Cahill, B., & Eberhart, J. (1978). Behavior of unacquainted saimiri troops upon encounter: A suggestive case study. *Primates, 19*, 643–656.

Erwin, J., & Flett, M. (1974). Responses of rhesus monkeys to reunion after long-term separation: Cross-sex pairings. *Psychological Reports, 35*, 171–174.

Fairbanks, L., McGuire, M., & Kerber, W. (1977). Sex and aggression during rhesus monkey group formation. *Aggressive Behavior, 3*, 241–249.

Fairbanks, L. A., McGuire, M. T., & Kerber, W. (1978). Effects of group size, composition, introduction technique, and cage apparatus on aggression during group formation in rhesus monkeys. *Psychological Reports, 42*, 327–333.

Gonzalez, C. A., Hennessey, M. B., & Levine, S. (1981). Subspecies differences in hormonal and behavioral responses after group formation in squirrel monkeys. *American Journal of Primatology, 1*, 439–452.

Goo, G. P., & Sassenrath, E. N. (1980). Persistent adrenocortical activation in female rhesus monkeys after new breeding group formation. *Journal of Medical Primatology, 9*, 325–334.

Hansen, E. W., Harlow, H. F., & Dodsworth, R. O. (1966). Reactions of rhesus monkeys to familiar and unfamiliar peers. *Journal of Comparative and Physiological Psychology, 61*, 274–279.

Jensen, G. D., Blanton, F. L., & Gribble, D. H. (1980). Older monkeys' (*Macaca radiata*) response to new group formation: Behavior, reproduction and mortality. *Experimental Gerontology, 15*, 399–406.

Joines, S. (1977). Introduction of incompatible gorillas (*Gorilla g, gorilla*) at the San Diego Wild Animal Park. *Animal Keepers Forum, 4*, 159.

Kaplan, J. R., Manuck, S. B., Clarkson, T. B., Lusso, F. M., Taub, D M., & Miller, E. W. (1983). Social stress and atherosclerosis in normocholesterolemic monkeys. *Science, 270*, 733–735.

Kaplan, J. R., Manuck, S. B., & Gatsonis, C. (1990). Heart rate and social status among male cynomolgus monkeys (*Macaca fascicularis*) housed in disrupted social groupings. *American Journal of Primatology, 21*, 175–188.

Kawai, M. (1960). A field experiment on the process of group formation in the Japanese monkey (*Macaca fuscata*) and the releasing of a group at Ohirayama. *Primates, 2*, 181–253.

Leshner, A. I., & Candland, D. K. (1972). Endocrine effects of grouping and dominance rank in squirrel monkeys. *Physiology and Behavior, 8*, 441–445.

Lindburg, D. G., & Robinson, P. (1984). Animal introductions: Some suggestions for easing the trauma. *Animal Keepers Forum, 11*, 222–225.

Mendoza, S. P., Coe, C. L., Lowe, E. L., & Serine, L. (1979). The physiological response to group formation in adult male squirrel monkeys. *Psychoneuroendocrinology, 3*, 221–229.

Moos, R., & Immelmann, K. (1987). The integration of an adult female gelada baboon (*Theropithecus gelada*). *Primates, 28*, 403–409.

Nash, L. T., & Flinn, L. (1978). Group formation in captive lesser galagos (*Galago senegalensis*). *Primates, 19*, 493–504.

Paluck, R. J., Lieff, J. O., & Esser, A. H. (1970). Formation and development of a group of *Hylobates lar*. *Primates, 11*, 185–194.

Reinhardt, V. (1988). Preliminary comments on pairing unfamiliar adult male rhesus monkeys for the purpose of environmental enrichment. *Laboratory Primate Newsletter, 27*, 1–3.

Reinhardt, V. (1989). Behavioral responses of unrelated adult male rhesus monkeys familiarized and paired for the purpose of environmental enrichment. *American Journal of Primatology, 17*, 243–248.

Reinhardt, V., Cowley, D., Eisele, S., Vertein, R., & Houser, D. (1987). Preliminary comments on pairing unfamiliar adult female rhesus monkeys for the purpose of environmental enrichment. *Laboratory Primate Newsletter, 26*, 5–8.

Reinhardt, V., Cowley, D., Eisele, S., Vertein, R., & Houser, D. (1988). Pairing compatible female rhesus monkeys for cage enrichment has no negative impact on body weight. *Laboratory Primate Newsletter, 27*, 13–15.

Reinhardt, V., Houser, D., & Eisele, S. (1989). Pairing previously singly caged monkeys does not interfere with common research protocols. *Laboratory Animal Science, 39*, 73–74.

Reinhardt, V., Houser, W. D., Eisele, S. G., & Champoux, M. (1987). Social enrichment of the environment with infants for singly caged adult rhesus monkeys. *Zoo Biology, 6*, 365–371.

Reinhardt, V., Houser, D., Eisele, S., Cowley, D., & Vertein, R. (1988). Behavioral responses of unrelated rhesus monkey females paired for the purpose of environmental enrichment. *American Journal of Primatology, 14*, 135–140.

Rhine, R. J., & Cox, R. L. (1989). How not to enlarge a stable group of stumptailed macaques (*Macaca arctoides*). In E. F. Segal (Ed.), *Housing, care, and psychological wellbeing of captive and laboratory primates* (pp. 255–269). Park Ridge, NJ: Noyes.

Rhine, R. J., & Kronenwetter, C. (1972). Interaction patterns of two newly formed groups of stumptail macaques (*Macaca arctoides*). *Primates, 13*, 19–34.

Rose, R. M., Bernstein, I. S., & Gordon, T. P. (1975). Consequences of social conflict on plasma testosterone levels in rhesus monkeys. *Psychosomatic Medicine, 37*, 50–61.

Rosenblum, L. A., Levy, E. J., & Kaufman, I. C. (1968). Social behaviour of squirrel monkeys and the reaction to strangers. *Animal Behaviour, 16*, 288–293.

Rosenson, L. M. (1973). Group formation in the captive greater bushbaby (*Galago crassicaudatus crassicaudatus*). *Animal Behaviour, 21*, 67–77.

Schildkraut, D. S. (1985). The introduction of peer-raised gorillas to a mother-infant pair. *American Association of Zoological Parks and Aquariums Annual Proceedings*, 679–683.

Scruton, D. M., & Herbert, J. (1972). The reaction of groups of captive talapoin monkeys to the introduction of male and female strangers of the same species. *Animal Behaviour, 20*, 463–473.

Southwick, C. H. (1967). An experimental study of intragroup agonistic behavior in rhesus monkeys (*Macaca mulatta*). *Behaviour, 28*, 182–209.

Strayer, F. F., Taylor, M., & Yanciw, P. (1975). Group composition effects on social behavior of captive squirrel monkeys. (*Saimiri sciureus*). *Primates, 16*, 253–260.

Wade, T. D. (1976). The effects of strangers on rhesus monkey groups. *Behaviour, 59*, 194–213.

Williams, L. E., & Abee, C. R. (1988). Aggression with mixed age-sex groups of Bolivian squirrel monkeys following single animal introductions and new group formation. *Zoo Biology, 7*, 139–145.

Williams, L. E., & Bernstein, I. S. (1983). Introduction and dominance manipulations involving old rhesus. *Folia Primatologica, 40*, 175–180.

Winston, L. A. (1985). Social interactions among female rhesus macaques (*Macaca mulatta*) during the nonbreeding season: Responses to the introduction of males. *American Journal of Primatology, 9*, 243–258.

Reintroductions of Infants to Mothers or Social Groups:

Alderman, T. (1986). Hands-off hand rearing of a gelada baboon, *Theropithecus gelada*. *American Association of Zoological Parks and Aquariums Regional Proceedings*, 679–683.

Devison, D. (1981). Notes on the early hand-rearing of an orangutan and its subsequent reintroduction to the mother. *Animal Keepers Forum, 8*, 89–90.

Dronzek, L. A., Savage, A., Snowdon, C. T., Whaling, C. S., & Ziegler, T. E. (1986). Technique for hand-rearing and reintroducing cotton-top tamarin infants. *Laboratory Animal Science, 36*, 243–247.

Gerlach, H. (1986). Socialization of artificially reared cotton-top tamarins in a peer-group and in an age-structured group. *Primate Report, 15*, 51–59.

Holman, S. D., & Goy, R. W. (1988). Responses of foster-mothers and troop members to adopted newborns in a captive group of rhesus monkeys. *Primates, 29*, 343–352.

Meder, A. (1985). Integration of hand-reared gorilla infants in a group. *Zoo Biology, 4*, 1–12.

Meder, A. (1990). Integration of hand-reared gorillas into breeding groups. *Zoo Biology, 9*, 157–164.

Mellen, J. (1978). Reintroducing an infant mandrill. *Animal Keepers Forum, 5*, 9–10.

Meyer, J. R. (1981). Reintroduction to the parents of a hand-reared lion-tailed macaque. *American Association of Zoological Parks and Aquariums Conference Proceedings*, 43–46.

Murdock, G. (1981). The reintroduction of Jessy, a hand-reared wooly monkey. *Animal Keepers Forum, 8*, 53–54.

Pook, A. G. (1974). The hand-rearing and reintroduction to its parents of a saddle-backed tamarin. *Dodo: Jersey Preservation Trust, 11*, 35–39.

Rohrer, M. A. (1977). Successful reintroduction. *Animal Keepers Forum, 4*, 133.

Visalberghi, E., & Riviello, M. C. (1987). The integration into a social group of a hand-reared brown capuchin. *International Zoo Yearbook, 26*, 232–236.

Watson, L. M., & Petto, A. J. (1988). Infant adoption and reintroduction in common marmosets (*Callithrix jacchus*). *Laboratory Primate Newsletter, 27*, 1–3.

Group Composition:

A comparison of the effects of individual cage housing along with discussions of optimal group size and age/sex class groupings.

Individual Cage Effects

Boot, R., Leussink, A. B., & Vlug, R. F. (1985). Influence of housing conditions on pregnancy outcome in cynomolgus monkeys (*Macaca fascicularis*). *Laboratory Animals, 19*, 42–47.

Brent, L., Lee, D. R., & Eichberg, J. W. (1989). The effects of single caging on chimpanzee behavior. *Laboratory Animal Science, 39*, 345–346.

Brinkman, C. (1986). "Psychological well-being" and the single social simian. *Australian Primatology, 1*, 1–2.

Brinkman, C. (1987). Single caging of social primates: Problems (and some solutions). *Australian Primatology, 2*, 17.

Castell, R., & Wilson, C. (1971). Influence of the environment on development of mother-infant interaction in pigtail monkeys. In H. Kummer (Ed.), *Behavior: Proceedings of the Third International Congress of Primatology, 3*, 121–124.

Cho, F., Suzuki, M., & Honjo, S. (1986). Adoption success under single-cage conditions by cynomolgus macaque mothers (*Macaca fascicularis*). *American Journal of Primatology, 10*, 119–124.

Davenport, R. K., Menzel, E. W., & Rogers, C. M. (1966). Effects of severe isolation on "normal" juvenile chimpanzees. *Archives of General Psychiatry, 14*, 134–138.

Floeter, M. K., & Greenough, W. T. (1979). Cerebellar plasticity: Modification of Purkinje cell structure by differential rearing in monkeys. *Science, 206*, 227–229.

Kessler, M. J., Turnquist, J. E., Pritzker, K. P. H., & London, W. T. (1986). Reduction of passive extension and radiographic evidence of degenerative knee joint diseases in cage-raised and free-ranging aged rhesus monkeys (*Macaca mulatta*). *Journal of Medical Primatology, 15*, 1–9.

Knezevich, M., & DeRousseau, C. J. (1985). Patterns of epiphyseal fusion in caged and free-ranging rhesus macaques. *Journal of Medical Primatology, 8*, 348.

Mitchell, G., & Gomber, J. (1976). Moving laboratory rhesus monkeys (*Macaca mulatta*) to unfamiliar home cages. *Primate, 17*, 543–546.

Reinhardt, V. (1990). Time budget of caged rhesus monkeys exposed to a companion, a PVC perch, and a piece of wood for an extended time. *American Journal of Primatology, 20*, 51–56.

Tarantino, S. J. (1970). Effect of cage confinement on social behavior in squirrel monkeys. *Psychonomic Science, 20*, 294–295.

Turnquist, J. E. (1981). Joint mobility in free-ranging versus laboratory *Erythrocebus patas*. *American Journal of Primatology, 1*, 326.

Turnquist, J. E. (1983). Influence of age, sex, and caging on joint mobility in the patas monkey (*Erythrocebus patas*). *American Journal of Physical Anthropology, 61*, 211–220.

Social Groups

Bernstein, I. S. (1966). An investigation of the organization of pigtail monkey groups through the use of challenges. *Primates, 7*, 471–480.

Bloomsmith, M. A. (1989). Interactions between adult male and immature captive chimpanzees: Implications for housing chimpanzees. *American Journal of Primatology, Suppl. 1*, 93–99.

Bowman, L. A., Dilley, S. R., & Keverne, E. B. (1978). Suppression of oestrogen-introduced surges by social subordination in talapoin monkeys. *Nature, 275*, 56–58.

Chalmers, N. R. (1971). Changes in mother-infant interaction following changes in group composition in sykes and patas monkeys. In H. Kummer (Ed.), *Behavior: Proceedings of the Third International Congress of Primatology, 3*, 116–120.

Dazey, J., Kuyk, K., Oswald, M., Martenson, J., & Erwin, J. (1977). Effects of group composition on agonistic behavior of captive pigtail macaques, *Macaca nemestrina*. *American Journal of Physical Anthropology, 46*, 73–76.

Fairbanks, L. (1974). An analysis of subgroup structure and process in a captive squirrel monkey (*Saimiri scuireus*) colony. *Folia Primatologica, 21*, 209–224.

Fitch, A. L., Merhalski, J. J., & Bloomsmith, M. A. (1989). Social housing for captive adult male chimpanzees: Comparing single-male and multi-male social groups. *American Journal of Primatology, Suppl. 1*, 87–92.

Hall, K. R. L., & Mayer, B. (1967). Social interactions in a group of captive patas monkeys (*Erythrocebus patas*). *Folia Primatologica, 15*, 213–236.

Hird, D. W., Henrickson, R. V., & Hendricks, A. G. (1975). Infant mortality in Macaca mulatta: Neonatal and postnatal mortality at the California Primate Research Center, 1968–1972. *Journal of Medical Primatology, 4*, 8–22.

Marsden, H. M. (1968). Behavior between two social groups of rhesus monkeys within two tunnel-connected enclosures. *Folia Primatologica, 8*, 240–246.

Marsden, H. M. (1973). Aggression within social groups of monkeys: Effect of contact between groups. *Animal Behaviour, 21*, 247–249.

Mason, W. A. (1975). Comparative studies of social behavior in callicebus and saimiri: Strength and specificity of attraction between male-female cage mates. *Folia Primatologica, 23*, 113–123.

McGrew, W. C. (1988). Parental division of infant caretaking varies with family composition in cotton-top tamarins. *Animal Behaviour, 36*, 285–310.

Oswald, M., & Erwin, J. (1976). Control of intragroup aggression by male pigtail monkeys (*Macaca nemestrina*). *Nature, 262*, 686–688.

Redman, H. C., & Schneider, R. (1979). An epidemiological study of neonatal and postnatal mortality in *Macaca radiata* at the California Primate Research Center (1966–1973). *Journal of Medical Primatology, 8*, 1–17.

Rhine, R. (1973). Variation and consistency in the social behavior of two groups of stumptail macaques (*Macaca arctoides*). *Primates, 14*, 21–36.

Ruppenthal, G. C., Harlow, M. K., Eisele, C. D., Harlow, H. F., & Suomi, S. J. (1974). Development of peer interactions of monkeys reared in a nuclear family environment. *Child Development, 45*, 670–682.

Sackett, D., Oswald, M., & Erwin, J. (1975). Aggression among captive female pigtail monkeys in all-female and harem groups. *Journal of Biological Psychology, 17*, 17–20.

Silk, J., Clark-Wheatley, C. B., Rodman, P. S., & Smauls, A. (1981). Differential reproductive success and facultative adjustment of sex ratios among captive bonnet macaques (*Macaca radiata*). *Animal Behaviour, 29*, 1106–1120.

Small, M. E. (1982). Reproductive failure in macaques. *American Journal of Primatology, 2*, 137–147.

Smith, E. O. (1973). A further description of the control role in pigtail macaques (*Macaca nemestrina*). *Primates, 14*, 413–420.

Stynes, A. J., Rosenblum, L. A., & Kaufman, I. C. (1968). The dominant male and behavior within heterospecific monkey groups. *Folia Primatologica, 9*, 123–134.

Suomi, S. J. (1974). Social interactions of monkeys reared in a nuclear family environment versus monkeys reared with mothers and peers. *Primates, 15*, 311–320.

Suomi, S. J., Eisele, C. D., Grady, S. A., & Tripp, R. L. (1973). Social preferences of monkeys reared in an enriched laboratory social environment. *Child Development, 44*, 451–460.

Vaitl, E. A. (1977). Social context as a structuring mechanism in captive groups of squirrel monkeys. (*Saimiri sciureus*). *Primates, 18*, 861–874.

Vaitl, E. A., Mason, W. A., Taub, D. M., & Anderson, C. O. (1978). Contrasting effects of living in heterosexual pairs and mixed groups on the structure of social attraction in squirrel monkeys (*Saimiri*). *Animal Behaviour, 26*, 358–367.

van Hoof, J. A. R. A. M., & de Waal, F. (1975). Aspects of an ethological analysis of polyadic agonistic interactions in a captive group of *Macaca fascicularis*. *Contemporary Primatology, Proceedings on the Fifth International Congress of Primatology*, 269–274.

Wolfheim, J. H., Jensen, G. D., & Bobbit, R. A. (1970). Effects of group environment on the mother-infant relationship in pigtailed monkeys (*Macaca nemestrina*). *Primates, 11*, 119–124.

Wooley, M. J., Clark, D. L., & Stevens, J. J. (1978). Squirrel monkey dominance and social behavior as related to group size and group structure. *Primates, 19*, 169–178.

Young, G. H., & Hankins, R. J. (1979). Infant behaviors in mother-reared and harem-reared baboons. (*Papio cynocephalus*). *Primates, 20*, 87–94.

Influence of Caretakers or Visitors:

Alvarez, F. (1975). Conditions of observation and social distance in groups of squirrel monkeys, *Saimiri sciureus*. *Primates, 16*, 465–470.

Asakura, S. (1957). A social relation between monkey and man—Monkey classifies men into high and low ranks. *Primates, 1*, 99–110.

Chamove, A. S., Hosey, G. R., & Schaetzel, P. (1988). Visitors excite primates in zoos. *Zoo Biology, 7*, 359–370.

Cooper, D. L., & Markowitz, H. (1979). Handler's effects on contact comfort behaviors of two trios of juvenile chimpanzees in the zoo. *Psychological Reports, 44*, 1015–1018.

Fa, J. E. (1989). Influence of people on the behavior of display primates. In E. F. Segal (Ed.), *Housing, care, and psychological wellbeing of captive and laboratory primates* (pp. 270–290). Park Ridge, NJ: Noyes.

Hosey, G., & Druck, P. L. (1987). The influence of zoo visitors on the behaviour of captive primates. *Applied Animal Behaviour Science, 18*, 19–29.

Thomsen, C. E. (1974). Eye contact by nonhuman primates toward a human observer. *Animal Behaviour, 22*, 144–149.

APPENDIX B

SPECIAL SESSIONS

The Conference on the Psychological Well-being of Captive Primates included a special poster and videotape session, the titles of which are listed below.

POSTER PRESENTATIONS

Alongi, C., Condon-Roybal, L., Williams, E., & Crowley, T. J. (1988). A low cost housing alternative for primates. (University of Colorado Health Sciences Center, Denver).

Boccia, M. L., Laudenslager, M. L., & Reite, M. (1988). Impact of different feeding regimens and dominance on behavior in socially housed bonnet and pigtail macaques. (University of Colorado Health Sciences Center, Denver).

Brent, L., Lee, D. R., & Eichberg, J. W. (1988). The effects on chimpanzee behavior of a physically and socially restricted environment. (Southwest Foundation for Biomedical Research, San Antonio).

Crockett, C. M., Sackett, G. P., Smith, O. A., Spelman, F. A., Morton, W. R., Ruppenthal, G. C., & Bowden, D. M. (1988). Space, exercise, and social needs of laboratory macaques. (Wisconsin Regional Primate Research Center, University of Washington, Seattle).

Desmond, T. (1988). Training voluntary cooperation with husbandry and research procedures in marine mammals. (Active Environments, Torrence, CA).

Desmond, T. (1988). Training to enhance socialization and reproduction in drills. (Active Environments, Torrance, CA).

Ernst, J., Evans, H. L., & Taylor, J. D. (1988). A brief test of foraging for socialization and evaluation of well-being: Comparison of macaques and tamarins. (Institute of Environmental Medicine, New York University Medical Center, New York).

Evans, H., & Taylor, J. D. (1988). Diurnal cycle in homecage behavior, a potential index of the well-being of laboratory primates. (Institute of Environmental Medicine, New York University Medical Center, New York).

Harrison, M. L., Sherwood, R. J., & Tardif, S. D. (1988). Differential use of wooden cage furniture in callitrichid primates: Implications for psychological well-being. (Department of Anthropology, Kent State University, Kent, OH, & Marmoset Research Center, Oak Ridge, TN).

Hayes, S. L. (1988). Increasing foraging opportunity for a group of captive cebus monkeys (*Cebus capucinus*). (Department of Zoology, University of Wisconsin, Madison, WI).

Hinds, K., & Ballard, R. (1988). Providing an enriched environment for rhesus monkeys in space flight. (NASA Ames Research Center, Moffett Field, CA).

Mahoney, C. J., & Moor-Jankowski, J. (1988). Criteria of successful methodology for psychosocial well-being of chimpanzees in a medical research laboratory. (LEMSIP, New York University School of Medicine, New York).

Maki, S., Fineg, I., & Bramblett, C. (1988). The psychological well-being of a captive guenon troop (*Cercopithecus aethiops* and *C. mitis*) enhanced by a naturalistic and enriched environment. (Department of Anthropology & Department of Animal Resources, University of Texas, Austin, TX).

O'Neill, P. (1988). Ten environmental models investigating short-term and long-term incidence of behavior disorders in laboratory-related rhesus monkeys. (National Institutes of Health Animal Center, Dickerson, MD).

Preilowski, B. (1988). Call for submissions. (Physiologische Psychologie, Rasthalde 3, D-7980, Revensburg, Weisenau, BDR).

Spector, M. R., & Bennett, B. T. (1988). The use of naturally occurring manipulanda can reduce the frequency of cage stereotypy in solitary-housed primates. (Biologic Resource Laboratory, University of Illinois, Chicago).

Tardif, S. D. (1988). The positive impact of biomedical research upon conservation of an endangered primate, the cotton-top tamarin. (Marmoset Research Center, Oak Ridge Associated Universities, Oak Ridge, TN).

Videotape Presentations

Johnson, L. (1988). Socialization and rearing conditions for cotton-top tarmarins at The New England Regional Primate Research Center. (Southborough, MA).

Reinhardt, V. (1988). Inanimate and animate enrichment at The Wisconsin Regional Primate Research Center. (Madison, WI).

Watson, L. (1988). Environmental enrichment for macaques at The New England Regional Primate Research Center. (Southborough, MA).

INDEX